# Cold War New England

## The Global Conflict at Home

By

Eamon McCarthy Earls

Cold War New England: The Global Conflict at Home



Via Appia Press

Franklin, Massachusetts 02038



viaappiapress.com

*And I have lived since--as you have--in a period of cold war, during which we have ensured by our achievements in the science and technology of destruction that a third act in this tragedy of war will result in the peace of extinction.*—Lester B. Pearson, Prime Minister of Canada, 1963-1968

Dedicated to Brigadier General John R. Mullin, US Army (ret.), an inspiration and himself a Cold War New Englander and to Michael, John, Joanne, Stephen and Caroline—this seems like the kind of eclectic book I would find on your shelves.

**Acknowledgements**

I wish to thank a number of different people for helping to make this book a reality. First and foremost, I wish to thank the archivists at the National Archives in Waltham, in particular Nathaniel Wiltzen, who furnished many of the best pictures for the book, along with the documents used to support the narrative. I would also like to thank my family, for their enduring support of my projects, as I forge ahead with yet another. Additionally, I'm grateful to the UMASS-Amherst Libraries for making available a plethora of digitized newspaper accounts of Cold War events in New England, without which it would have been impossible to tell this story.

## Introduction: On the Alert

The first blow of the war came almost unexpectedly at 1:22 in the afternoon on June 15, 1955. New Bedford on Massachusetts' South Coast was the first American city to experience the terrifying reality of nuclear war. A single H-bomb dropped from a Soviet bomber leveled much of the small fishing and manufacturing port. Within minutes, Soviet bombers were over Boston and Providence, delivering equally devastating blows to two of New England's proudest cities.

Across the country, the story was similar. Soviet bombs fell on Washington, DC, New York City, Philadelphia, Chicago and the major Californian cities, reducing each target to rubble and sending clouds of radioactive fallout into the stratosphere. Fortunately, the US government had enough foresight to send the president underground, and evacuated 15,000 federal workers from the major departments to secret headquarters in Virginia and Maryland.

Less than ten years after the bombing of Hiroshima and Nagasaki most major cities in the US lay in ruins. Over five million people were dead, another five million were seriously wounded, and more than ten million were left homeless, with fallout spreading rapidly to surrounding areas. This unthinkable attack on America actually happened, although it was only an exercise in civil defense preparedness. The largest single nuclear alert that the US has ever experienced encompassed cities and towns across the nation and was taken very seriously by authorities at all levels. Although the attack was a fiction,

President Eisenhower and much of the federal workforce evacuated the capital as they would in advance of a real crisis.

In New York City, the nation's biggest city, Operation Alert was treated as an actual civil defense emergency, with officials responding accordingly. Streets were shut down and New York residents were sent underground into the subways and fallout shelters. The police arrested 30 people for noncompliance in New York City alone—mainly beatniks and peace activists, but also unruly drivers and pedestrians.

In other cities, the drill was taken with a much larger grain of salt. Philadelphia largely ignored the drill, along with Peoria, Illinois, a city that refused to play along with the deadly "fallout cloud" drifting in their direction from Chicago. Even in Washington, DC, not everyone took the drill seriously. Many federal workers stayed at their offices and children waved at "Ike" as his limo powered down the capital's wide boulevards. Operation Alert was described by one Civil Defense Administration official as, "[not a drill] but a show."

While Operation Alert definitely tended towards theatrics, it made perfect sense in 1955. Today, Americans would probably ignore even much more local calls to action in the name of civil defense or national preparedness. However, each time we board a flight we gladly consent to the barrage of pat downs, x-ray machines and metal detectors that have become the leading tools against hijacking and terrorism. Threats may have changed, but Americans are still scared of an unpredictable world. In the early 21st century, Cold War artifacts seem incredibly archaic—whether tacky 'duck and cover' newsreels from the 1950s, footage of LBJ, Nixon, and Vietnam, or the rusting fallout shelter signs that still decorate the walls of many public buildings. Today, the Cold War is old news; that chapter closed almost 25 years ago when crowds of Germans appearing in muddy video footage knocked down a wall dividing Berlin in two.

The Cold War may be old news to most of us, but we cannot forget how recently it ended and its enormous long term impact on the US. Even as news agencies have recently covered the removal of troops in Afghanistan, a conflict dubbed America's longest war, we forget about the Cold War, a conflict that lasted for more than 45 years, costing the US trillions of dollars and the lives of more than 100,000 service members.

Many things have been written about the Cold War—accounts of egotistical generals and leaders, of pivotal events, and theatres of the war such as Vietnam and Korea. But what about the impact that the Cold War had on the home front? How did America's longest war impact a single region—New England—and how did this single region impact the final outcome of this undeclared war?

New England has always been one of the most productive regions in the US. Densely populated and situated closer to Europe than any other part of the country, New England was positioned to play a pivotal role when the Cold War began in 1945. Between the '40s and the '90s, tens of thousands of New England service members served at home and abroad, defending the US. Meanwhile tens of thousands of others worked on the home front developing and manufacturing brand new technologies, ranging from radar and computers, to missiles and planes. New England was truly America's forward base at home.

Few today could guess that Vermont's rolling, verdant landscape once concealed ICBM silos, that the world's largest air base was once located in Maine, that cities such as Boston and Providence were ringed with nuclear-tipped air defense missiles, or that American fishing boats waged a 20 year war against offshore Soviet factory ships. New Englanders stood up in the face of terrifying new threats, serving in the military, innovating at companies and universities to create some of the first and most advanced computers and weapons systems of the time. The story of Cold War New England is a story of people, of industries, of bases, and of new technologies with incredible power and sophistication. I hope that you find New England's strange "wartime" history as interesting as I do.

--Eamon McCarthy Earls, 2016

## Chapter I: The Fragile Peace

In 1948, trouble was looming. Communists, backed by the Soviet Union seized control of the Czechoslovakian government. A mere three years after the end of World War II in Europe, it looked as if the entire continent might erupt into another, even deadlier squabble among the Allies themselves. This possibility became apparent in June, 1948, when Soviet soldiers severed the road, rail and canal connections between Western Germany and West Berlin. Under the terms of occupation, Germany and Austria had been divided between the Allies. The French, British and Americans controlled several zones in the west, a region that would later become the Federal Republic of Germany, better known as West Germany. The Soviets controlled the east, a region that would soon become the totalitarian German Democratic Republic, known to most people as East Germany. Austria was ultimately spared from the divisions between east and west, by declaring independence as a neutral nation in 1955.

Berlin, the capital of the former Nazi Germany (located in the Soviet-controlled eastern part of Germany) had been divided on a much smaller scale by the Allies. The same pattern held true for Berlin. The western allies controlled West Berlin, while the Soviets controlled East Berlin (the then future capital of East Germany).

To counter the Soviet blockade, the Americans, British and French launched a daring around the clock airlift—the most extensive ever conducted in history—to supply West Berlin with all of its needs. Food, medicine, clothing and even coal were shipped in by supply planes arriving at Tempelhof airport. Berlin had been virtually leveled in the war, and hundreds of thousands of citizens eked out a meager existence in the midst of the ruined city's rubble. As they flew over the city, American pilots threw fistfuls of candy to German children below.

In spite of the tremendous need for transportation in Europe, 1948 saw the closing of Presque Isle airbase in northern Maine. The base, on the outskirts of the small city of Presque Isle had recently been re-designated Presque Isle Air Force Base with the creation of the Air Force in 1947. On February 20, the base was officially inactivated, except for a few maintenance workers left behind to keep the runway free of ice. It was a surprising move, considering the base's importance during World War II. Between 1942 and 1944, thousands of Air Transport Command flights had left the runway for airfields in Europe as part of the vital trans-Atlantic air supply route for the Allies. The Air Force cited the construction of nearby Limestone Air Force Base, in Limestone, Maine, slightly to the northeast of Presque Isle as the reason for the inactivation.[1]

With bad news from Europe becoming ever more common, the military took the opportunity to instill some public confidence on St. Patrick's day, in 1948. While proud Boston-Irish bands traipsed down the streets of South Boston on March 17, Army and Navy aircraft roared by overhead.

Air Force planes joined the display. Propeller driven fighter planes, such as P-48s, P-47s, and P-61s rumbled in from Stewart and Dow AFB in New York and Bangor, Maine. Some aircraft arrived from closer to home, flying in from Grenier field in New Hampshire and Hanscom AFB in Bedford.

The Air Force was joined in the sky by over 80 planes from the Navy's aircraft carrier, *USS Lexington*. The air armada was led by local pilots from communities such as Lexington, Clinton, West Roxbury, Norwalk, Malden and Waltham.[2]

While air units led the fight to resupply West Berlin, soldiers at Fort Devens braved the muggy, midsummer heat in Central Massachusetts, fighting a mock battle with the 3rd Army. Soldiers from the 3rd Army attempted to fight their way across the large military reservation, 'battling' the opposing forces of local Devens' troops. The exercise had many similarities to actual combat. Mosquitoes and several rainstorms added to the intense summer heat, in combination with thick underbrush--likely rich in brambles and poison ivy--and water rationing across the base. Soldiers were issued special, highly recognizable uniforms to prevent them from switching sides. Base engineers practiced building bridges, which were quickly destroyed with enemy explosives. Meanwhile, fighter planes from the 82nd Fighter Wing at Grenier arrived to strafe the different forces, while soldiers struggled to hold their lines by laying minefields. The exercise was brought to an end, "by the withering machine gun fire, harassing mosquito attacks and the demoralizing effect of C rations," according to the *Boston Globe*.[3]

The changing international climate led the Air Force to switch Grenier AFB's role to air defense on August 24, 1949. The Manchester, New Hampshire base was transferred from the Strategic Air Command, the new Air Force bomber command based in Omaha, Nebraska. Grenier now fell within the Continental Air Command, based in upstate New York. "Local Air Force units will be a part of the defensive force to protect the United

States instead of a striking force against an aggressor in home base territory," the *Boston Globe* explained.[4]

A few days later, Defense Secretary Louis Johnson and the Senate trimmed the Air Force budget. In addition to trimming away 18,000 civilian jobs and hundreds of millions of dollars in contracts, the Senate agreed to inactivate six fighter groups around the country. At Maine's Dow AFB in Bangor, the 14th Fighter Wing was ordered to pack their bags in preparation for inactivation together with the 82nd Fighter Wing at Grenier. The 2000 personnel of the 14th Fighter Wing were late in getting official word about their reassignment, as they waited around at the base where they had lived and worked since October, 1946.[5]

It seemed that in spite of European troubles, the federal government viewed any threats as distant. After all, there was over 3,000 miles between the American East Coast and the nearest Soviet soldiers. The vast Red Army, cobbled together with millions of peasants did not seem like the kind of threat that would require fighter planes to counter.

Such attitudes changed two days later, on August 29, when the Soviet Union detonated a small nuclear bomb on the Kazakh steppe. The Soviet development of an atomic bomb in 1949 terrified and surprised the US. A new chapter was opening in the nascent Cold War and the threat of conflict seemed to be growing too soon after the promises of peace made at the end of World War II.

In spite of the revelations about the Soviet's and their new armaments, the Air Force proceeded with its closure of Presque Isle AFB in October. On October 4th, the Air Force announced that it was selling the 23 million dollar base to City Manager Norman

MacDonald for one dollar. MacDonald met with the Air Force to secure a deed for the 6,000 foot runway and its surrounding buildings, which he was eager to see used by the local company, Northeast Airlines. In a meeting with the city council, MacDonald assured the council that purchasing the base was a "good deal" that would offer, "unlimited opportunities." At just one dollar, it was truly an offer that Presque Isle could not pass up![6]

In 1950, tensions were further escalated when North Korean troops poured into South Korea. Almost overnight, the US found itself fighting a losing war in the inhospitable terrain of the Korean Peninsula. The initially small US force, left over from the occupation of the peninsula after the surrender of Japan, was pushed back to the southern city of Pusan, along with the remnants of the inexperienced South Korean army. When fresh US troops arrived, along with reinforcements from other UN nations, the tide turned quickly, with the US-UN forces pushing the North Koreans to the northern end of the peninsula at the Yalu River in only a few months.

Thousands of men and women from New England were mobilized to fight in Korea, many shipping out at a moment's notice to fight the North Koreans. Unfortunately for the UN force, the push to the Yalu was deemed a serious threat by the fledgling People's Republic of China, which began pouring tens of thousands of Chinese "volunteers," into North Korea under Chairman Mao's direction. The Korean War see-sawed back and forth across the peninsula, finally concluding in 1953, with the pre-war borders of North and South Korea hardly changed (although a cease fire was instituted, the Korean War officially continues to this day).

Back in New England, summer exercises at regional bases took on a new found significance for everyone from infantrymen to pilots.

Pilots flying with the Massachusetts and New Hampshire Air National Guard spent the start of August, 1950, training at Grenier and Barnes Airport, in Westfield, Massachusetts aboard newly acquired and recently re-designated F-51 Mustang fighters. The shiny aluminum wings and fuselages of Mustang's had glinted in the skies over Europe five years earlier as American bomber crews turned Germany into a "fortress without a roof." The Mustangs served as long-range fighter escorts for lumbering B-17 and B-24 bombers, helping to fight off German flyers sent to intercept the bombers.

The Massachusetts pilots engaged in a two week training period at Grenier, during which they received the General Carl Spaatz trophy. The award was significant, speaking of a low accident rate and highly successful training. The 130 officers and 600 airmen involved in the training were acquainting themselves with every aspect of the F-51s, a challenge for everyone, including the mechanics.

Although the aircraft themselves were probably somewhat familiar to many of the men at Grenier, from prior service in World War II, their mission was dramatically different. Exercises at Grenier simulated interceptions of enemy aircraft, ranging over large areas of New England and eastern Canada, instead of escorting 'friendlies' through enemy skies. Bringing outside aircrews into an unfamiliar base was good preparation for the kind of expeditionary missions that the men might expect to face in Korea or Europe.

A *Boston Globe* reporter explained the changing circumstances for aircrews at Grenier. "The A-bomb, faster jet fighters, and larger, longer range bombers were given by the [Air Force] spokesman as the reason for the new training plan." In fact, the F-51s, which had been on the cutting edge only a few years earlier, were outmatched in the Massachusetts and New Hampshire skies by F-84 Thunderjets from the 101st Fighter Squadron at Otis AFB on Cape Cod. The F-84 was America's first generation jet fighter. Although it was somewhat primitive compared to modern jet fighters, it remained leagues ahead of the F-51s. Most significantly, the Otis F-84s could outpace the F-51s, and in some cases may have even been more durable than the F-51s, which relied on fragile piston engines with a proclivity for unexpected malfunctions. The 101st Fighter Squadron's air units were joined on the ground by weather and maintenance wings.[7]

While pilots, air crews and meteorologists trained at Barnes, Grenier and Otis, 150 Reserve officers poured into Boston on August 12, for two weeks of intense summer training. Many of the officers were leading veterans and business executives in the city who didn't have far to travel to get to their posts. According to the *Boston Globe*, the men, "will be ready to take over control of the port of Boston should it have to be activated as a military port of embarkation." Technically known as the 9240th Organized Reserve Technical Unit, the group was more informally known as the Boston Port of Embarkation. Unusually for an Army unit, it was made up almost entirely of upper level, commissioned officers, accompanied by 35 rank and file enlisted men from the 356th Port Battalion.

Jointly led by Col. Henry Trask and Col. James Cunningham, from Milton and Medford respectively, the group insisted that the training had nothing to do with the partial mobilization for Korea. Both commanding officers were heavily involved outside of their Army commitments. For instance, Trask served as the treasurer for Trask Engineering Corporation, a group responsible for building, among other things, the Derby Racer roller coaster at Revere Beach.

The Army group had rehearsed port management for the first time a year earlier at Camp Kilmer, New Jersey. At the time, the post in New Jersey was the only officially mobilized port in the nation.

During one week of the training, the soldiers toured New York's harbor, and trained alongside fellow officers in the Port of New York. Although most of the trainees were men, they were joined by 10 women from the Women's Auxiliary Corps (WAC).[8]

On August 15, while the Army officers rehearsed in New York's harbor, thousands of New Englanders gathered at the Eastern States Exhibition in West Springfield to see, "displays ranging from baking and candle dipping to atomic energy." The Exhibition, a perpetually popular occasion, had attracted 400,000 visitors in 1949. The Navy set up a display of sonar gear from the Naval Submarine Base in Groton, Connecticut, alongside, a small State Board of Fisheries aquarium, Thompsonville rug-making machines, and jet engines built in the Nutmeg State by Pratt & Whitney.[9]

Throughout the final days of August, 1950, the Guiel family in South Hadley wrung their hands together with thousands of others throughout the country. Although only weeks old, the Korean War was turning swiftly bloodier, and their 20 year old son, Alfred, was listed among the missing. A telegram from the Department of Defense had launched 10 days of incredible anxiety, interrupted by occasional bits of good news. A ham radio operator on Guam relayed a message to the family, telling them that their son was actually safe. The Guiel's hardly dared to believe the message.

Definitive good news finally came on August 31, when Alfred arrived at Westover AFB in Chicopee, close to the Guiel's South Hadley home. Alfred had sustained injuries and was dispatched to Chelsea Naval Hospital in Boston to recuperate—but at least he was home safe.[10]

Two days later, less encouraging news reached the family of a Navy pilot from New Durham, New Hampshire. Lt. Robert W. Haller was flying a Corsair fighter plane out of Squantum Naval Air Station in Quincy, on what was supposed to be a mock 'dive-bombing' run of Otis, but none of the other pilots involved with the mission spotted his plane after it left the base. While pilots from Westover, Squantum and Otis raked the ground with their eyes, in search of Haller's Corsair, Navy officials reported that he might have flown over Boston mistakenly with a separate flight to 'dive bomb' New Hampshire's Isle of Shoals. From there, it was presumed that his plane might have veered over New Hampshire and crashed. Indeed, the wreckage was eventually spotted in the Granite State.

The flight crews leaving from Squantum to take part in the search were among the last to fly out of the base. On November 10, 1950, 20 Navy and Marine squadrons were told to get ready for a move the nearby South Weymouth NAS, to alleviate overcrowding in the Boston skies. Squantum would be closed down soon thereafter. The move occasioned a great deal of construction in South Weymouth including new hangars and base offices for hundreds of fresh personnel. Construction workers also had to figure out how to build a runway in a hurry. South Weymouth had been used for blimps throughout World War II and into the early years of the Cold War, but had managed to get by without functioning runways as a result.[11]

With the new Soviet atom bomb, victory by the "Reds" in China, and then the outbreak of the Korean War, the US began to take the threat of a nuclear war very seriously. In response, the nation quickly ramped up its bomber force, in case a strike on the Soviet Union was needed, and likewise beefed up its air defense forces substantially. New England fighter pilots who had felt underappreciated as recently as 1950 were suddenly back in the thick of the action.

On September 23, 1951, fighters from South Weymouth raced skyward from newly constructed runways to intercept 'enemy' planes arriving from New York. 'Operation Clambake' was accompanied by an air show at Hanscom. Taken together, they crowded up the Boston skies for a day.[12] 'Operation Clambake' inaugurated South Weymouth as a modern base. In June, a quarter ton of dynamite had leveled one of the four million dollar blimp hangars built during World War II, to make way for the new runway.[13]

Soon after the exercise at South Weymouth, the Military Air Transport Service abandoned Westover in favor of New Jersey's McGuire AFB. For a brief instant, Westover's future seemed uncertain.[14] Meanwhile, in the community of Newington, New Hampshire, locals—and their neighbors in 13 other towns—rallied against a 46 million dollar bomber base. "We feel Washington should find out the truth," said Selectman Newton Weeks. "We are loyal Americans, but as citizens we want to know why thousands of acres of land already owned by the government only a few miles away [in Sanford, Maine] cannot adequately meet the needs of Air Force." The Air Force intended to take local farms and house lots to convert into national security property.[15]

The proposed air base in Newington was likely sited in that community for its proximity to the Portsmouth Naval Shipyard, and the New England coastline. Portsmouth was already accustomed to working with service members. Up and down the coast, bases came roaring back to life. April 1, 1953 marked the move from Squantum to South Weymouth. Even before that date, bases had plenty of reasons to be active. On September 12, 1952, the Naval Public Information Office in Boston announced that foreign submarines (presumably from the Soviet Union) had been sighted off the coast by aircraft operating out of Quonset Point. Around the same time, the Navy and the FBI collaborated on a sabotage investigation at Charlestown Navy Yard. A person, or persons, unknown had intentionally ruined the reduction gears on the destroyer *USS Pritchett*, shortly before a training voyage. Damages ran to 300,000 dollars.[16]

Perhaps with such revelations in mind, Quonset Point launched Operation Codfish on September 20. A fictional aircraft carrier group 'attacked' New England in the night, and slipped away from the 250 planes pursuing it. Now in its final days, Squantum became the center for operations, welcoming planes from Quonset Point and Brunswick NAS for touch and go stops as they chased the 'fleet' between the tip of Cape Cod and Nantucket.[17]

Exercises like Codfish and the training at South Weymouth helped to prepare some pilots for deployment. A squadron of F-86 fighters stopped briefly at Dow AFB, in Bangor, Maine on their way to cross the Atlantic. There, they would join patrols along the Iron Curtain, flying out of West German bases.[18]

In April, 1955, the new Strategic Air Command (SAC) a major Air Force command in charge of the US strategic bombing force, took over Westover AFB, turning the transportation hub into a combat ready installation. But before Westover could accommodate SAC bombers, modifications needed to be made. A new 300 foot wide, 11,000 foot long runway was built along with "nosedock," shelters for KC-135 air refueling tankers. The KC-135s were among the nation's first jet tankers, able to keep pace with the roaring turbojets that propelled B-52 Stratofortresses.

While the construction of a new runway took place, secret construction was underway at other areas of the base. Three rings of electrified fences were erected surreptitiously in Ludlow, forming the frontier of the highly classified Stony Brook Air Force Station. The new station, was a storage and repair facility for nuclear weapons, the largest such facility in the world. Ludlow, Massachusetts, a sleepy Western Massachusetts town hosted thousands of nuclear warheads at its peak.

Almost as secret as Stony Brook was P-1900, a photographic analysis division built at Westover during the '50s building boom. When it became operational, P-1900 helped to track Soviet troop and ship movements, as members of the 8th Reconaissance Technical Squadron worked to create accurate maps of the USSR, Eastern Europe, and even US allies. Initially, photographs were acquired through aerial reconnaissance by U-2 spy planes, although later on in the '60s, photos came from first generation spy satellites, with film dropped from space.

Photo analysts at P-1900 were heavily supervised, often joking that they were supervised even using the bathroom. Those who served at P-1900 never left, while on duty. The facility had its own medical staff, and a crew of 10 people in charge of the air conditioning system, built to deal with strong fumes from photographic development chemicals.

In 1956, the base added brand new B-52 Stratofortress bombers, complementing the lumbering B-36 Peacemaker bombers that were already in use at Westover. The B-36 was the largest bomber ever built for the US arsenal. Propelled by six "pusher" propellers, a single B-36 could make houses and even the ground shake when it flew overhead.

In between missions, when they were off duty, flight crews and the thousands of other Air Force staffs at Westover flocked to the Red Barn, an old barn converted into a bar. Complete with jukeboxes, the Red Barn was the ideal hangout for Air Force personnel after a long day of nuclear deterrence!

Changes at Westover also brought changes to Chicopee. Students began to appear at Chicopee schools who had been born at far flung bases in Libya, Turkey, Britain and France.

Almost 500 miles northeast of Westover, on the far eastern flank of the US, another mega-base was taking shape in Maine potato country. Aroostook County, the northernmost county in Maine, and the largest county east of the Mississippi was the leading producer of potatoes in the US at the end of the '40s, and a great place to find open land, with its sparsely populated rural landscape. Planning and construction of Limestone AFB began in 1948, two miles to the northwest of the small town of Limestone. The 8900 acre Air Force property was situated on some of the flattest land in New England, other than the Pioneer Valley in Western Massachusetts (where Westover was situated), roughly 600 feet above sea level and only five miles west of the Canadian border with New Brunswick.

Limestone, the product of significant government investment, became active in 1953. Situated closer to Europe, the USSR, and the Middle East than any other base in the continental US, Limestone was perfectly located to become America's advance base for launching or countering operations to the east. Even in its planning stages, the base was singled out as a staging area for the enormous B-36 bombers that represented the forefront of US strategic potential. With plans for up to 100 B-36s to be based at Limestone, Air

Force builders spent millions constructing 10 huge hangars, each capable of accommodating two B-36s. The initial order proved insufficient, and more hangars to accommodate an additional 10 ten bombers were added.

February 25, 1953 marked the activation of the 42nd Bombardment Wing, flying B-36s for SAC. Aircraft at Loring (as the base was soon renamed) were updated in 1956 to include brand new B-52C Stratofortresses, jet bombers.

Among the 110 SAC bases ultimately opened across the US, and the nearly 50 other bases overseas, Loring and Westover were "flagship" installations. But there *were* other sites of SAC activity in New England, notably Presque Isle AFB, Grenier AFB, and Pease AFB.

The flipside of the SAC coin was the Aerospace (Air) Defense Command, a command tasked with protecting the US against enemy air attack. At some of the smaller Air Force bases in New England, the '50s were all about air defense activity.

At Presque Isle, pilots of the 74th Fighter-Interceptor Squadron flew F-86 and F-94 jet fighters over the chilly northern frontier of the continental US between 1951 and 1954. In '54, the men of the 74th were shipped out to northern Greenland, a much sleepier posting than Presque Isle!

Picking up the slack at Presque Isle, the 75th and 76th Fighter-Interceptor Squadrons continued to fly F-86 fighters through the mid-'50s. The interceptor squadrons at Presque Isle rotated frequently. Units would come and go every few years, with some squadrons sent to upstate New York and others to Keflavik field in Iceland.

Dow AFB in Bangor was home to the 49th Interceptor Squadron and the 174th Fighter Squadron. The fighter groups at Dow faced challenges similar to their comrades at Loring and Presque Isle. During the winter months, the men would compete with other air bases around the globe to see who was having the worst weather. Dow was usually pretty bad on the weather front, receiving two to three feet at a time, and experiencing drifts of up to 10 feet that made runways impassable. Not surprisingly, though, Thule, Greenland still won the weather contest.

New Hampshire's twin air bases, Pease (outside of Portsmouth) and Grenier (outside of Manchester) hosted fighters, transports and bombers. Grenier was used by the "second line" fighters kept around as backup for the F-84s and F-86s elsewhere.

On October 4, 1957, the USSR launched the first artificial satellite into orbit. Sputnik 1 orbited the Earth for three months, emitting radio chirps picked up by ham radio operators across the globe. The launch of the first satellite took the US government by surprise.

Although it was a scientific embarrassment for America not to have achieved this space superlative, the implications of Sputnik were much more significant, suggesting that the Soviet Union could bombard the US from afar with long-range missile. Before Sputnik, during the mid-'50s, the US was terrified of Soviet bombers being sent to attack the US, coming across the Atlantic from East Germany or flying over the Arctic to strike US and Canadian cities.

The Pine Tree Line was the first of the radar early warning systems set up by the North American Air Defense Command. Spanning southern Canada, radar installations run primarily by the Royal Canadian Air Force were set to warn the government of incoming Soviet bombers. Small numbers of "Pine Tree" stations were run by the US at air bases in Newfoundland, administered from New England. The Pine Tree Line was soon replaced by the Mid-Canada line, which was also rapidly overturned by the Distant Early Warning (DEW) line above the Arctic Circle.

The DEW line would serve as an early warning network into the '90s, and was designed and built through the cooperation of US and Canadian governments. Curiously, the DEW line concept was first proposed in 1952 by a group of MIT scientists part of the Summer Study Group. As it unfolded, DEW was not the only air defense system dreamed up in Massachusetts.

At a time when radar was still a new technology, MIT's Lincoln Laboratory created a proposal in 1952 to borrow a page from the oil drilling industry and set up forward radar stations at sea on platforms modeled after oil rigs. Taking the name "Texas Towers" after the numerous oil rigs along the Texas coasts, three such structures were built off the Atlantic Seaboard.

One tower, scheduled to be built off of New Hampshire was cancelled but two towers were built off the coast of Long Island, with another constructed in the Georges

Bank off of Massachusetts. Texas Tower 2, run by the 762nd Radar Squadron, was attached to North Truro Air Force Station on Cape Cod. Built in the late '50s it was decommissioned only a few years later in 1963.[19]

The Texas Towers system was resupplied from Massachusetts, with helicopters from Otis AFB bringing frequent deliveries, and larger loads traveling aboard the *USNS New Bedford*, based out of New Bedford, Massachusetts.

When the Texas Tower system was ultimately shut down in the '60s, early warning along the Atlantic Seaboard took to the skies. The mission of the Texas Towers was taken over by modified Lockheed Constellations flying from Otis. The cigar shaped "Connies," redubbed EC-121 Warning Stars would operate until the late '70s.

The Texas Tower system, and the entire air defense system for the country was linked up to the Semi-Automated Ground Environment (SAGE) system from 1958 through 1983. The SAGE system was decades ahead of its time. Attempting to better prepare air defenses to deal with an attack, the US and Canadian governments created the most advanced computer system of its day.

The origins of SAGE, like the DEW line, lay in Massachusetts. The program started with the Cape Cod System, set up in the early '50s by MIT and the Air Force. Using agile B-47s as "stand-ins" for the Soviets, radar outposts fed digital information over the phone lines to the Whirlwind I computer in Cambridge, which was fast enough to warn fighters of the "attack."

Ironically, the millions of dollars the US spent investing in Sky Watcher programs, the Pine Tree Line, and Texas Towers, the US was much better positioned to wage an air war against the USSR. B-36, B-52, and B-47 bombers could actually reach the Soviet Union, whereas the Soviets were just introducing the first of their long range 'Bear' bombers. If the Soviets had launched an attack, many of their aircrews would have had to fly suicide missions, with no hope of return due to lack of range.

Sputnik pushed the US space program into high gear, and led the Air Force to speed up its development of first generation intercontinental ballistic missiles (ICBMs) and air-breathing Snark cruise missiles. One of the first and most advanced ICBMs in the Air Force arsenal was the Atlas. Work on the missile began in 1953 and the first missiles came into production in 1959. US fear of Soviet missile attack, and demands for retaliation would lead to the installation of Atlas missiles in the early '60s, in western New England.

In 1957, the Strategic Missile Site Selection Panel chose Presque Isle as the site for the first Snark missile base. By 1959, doubts were raised about the Snark, leading the Air Force to deploy only one Snark squadron. That year, the 702nd Missile Wing was created at Presque Isle, and added to the sprawling 8th Air Force, as the first missile wing of its type. The SM-62 Snark, was a ground launched cruise missile capable of traveling almost 6,000 miles. Unfortunately for crews, the Snark had a baffling navigation system, and was never widely deployed, with only 30 placed in service. Presque Isle would be the only base to ever host the missile.

As the Air Force dreamed of wars spanning continents and oceans, the Army was busily carrying out the "grunt" work that infantrymen are used to. Ft. Devens was almost cast off by the Army in the immediate aftermath of World War II, with a sudden drop in troop numbers at the end of the war. When the Korean War broke out, the base went back to doing what it was used to, staging and preparing soldiers for combat. Numbers of troops shot up, rising almost to World War II levels. Fort Devens probably seemed like the last place that would be swept up in the "Atomic Age," but in 1958, the Atomic Energy Commission suggested that some of the swathes of open land at Devens be used to build a nuclear processing facility. Fortunately the people of Ayer and Shirley were spared this particularly environmentally destructive project.

Along New England's coast, the Navy sailors were almost as busy as their brethren soldiers and airmen inland. The US Navy had finished World War II as the largest fleet in history, scrapping many of its supply ships in the immediate aftermath of the war. When the Cold War began, the Navy began expanding the fleet once again, reaching a total of 1,030 ships in 1955.

Boston Navy Yard in Charlestown had churned out destroyers and destroyer escorts during the '40s, and continued building on a more limited basis throughout the '50s. In fact, two *John C. Butler*-class destroyer escorts, *USS Wagner* and *USS Vandivier* had both been built and launched in the '40s, then "mothballed" after the peace. Both ships returned to their birthplace in Boston in 1955 to be refitted as radar pickets.

A short distance down the coast from Boston, Bethlehem Steel Corporation's Fore River shipyard was finding new business for itself in the Cold War. *USS Northampton*, a 13,700 ton cruiser laid down in the 1944 was finally launched in 1951 and commissioned in 1953. In the early '60s, at the height of the Cold War, *USS Northampton* was refitted to become part of the National Emergency Command Post Afloat. Designed to be a naval version of Air Force One, "a floating White House," Northampton was given a bizarre array of antennae and dishes to communicate with the country in times of war.

Not far past Quincy, Hingham Naval Ammunition Depot accounted for one of the largest naval presences in New England. The depot had seen a drop in activity during the '50s, but still bragged 90 buildings, its own telephone interchange and 15 cranes.

In nearby Cohasset, the Hingham Naval Ammunition Depot Annex stretched across seven square miles of land, into Norwell and Scituate. The annex may be the only area of landmines in Massachusetts. During the '50s, the site's many bunkers were used to store rocket motors and nuclear depth charges. In fact, according to historian Jim Rose, Bunker N9, the only one of the 108 bunkers at the site preserved today was used to store RUR-5 ASROC missiles throughout its service life. An acronym for 'anti-submarine rocket,' ASROC was common in eight-cell missile batteries on US warships, designed to take down undersea Soviet foes with acoustic torpedoes or a W-44 nuclear warhead. The ASROCs and their attendant warheads were transported to Cohasset from Quonset Point in a six-vehicle column, guarded by Marines, and then put on a 'lighter' barge and carried to an offshore warship in Boston harbor.[20]

Both sites were closed in 1962, and sold between 1967 and the early '70s to become Bare Cove Park (Hingham) and Wompatuck State Park, although unexploded shells and other ordnance remained a long-term problem in some parts of the park.

Activity at the annex was gradually scaled back and the base was ultimately closed, although an Army Reserve infantry brigade stayed active at the site until 1982. Most of the bunkers at Cohasset were ultimately removed, with only a few left preserved.

After the rapid escalation of the Cold War during the early '50s, New England's line-up of bases, factories and laboratories began to settle into a routine that would fluctuate in response to Soviet blunders, American defense budgets and political vagaries over the next 40 years.

The late '50s were marred by more horrific crashes, continuing a trend that would last throughout the Cold War years. On June 27, 1958, a brand new KC-135 jet tanker with 15 crewmembers crashed alongside the newly built Massachusetts Turnpike, exploding and killing all hands onboard. In addition to the aircraft's normal crew were six journalists from groups such as *Time*, United Press, the *Boston Herald*, and *US News & World Report*.

Chicopee police patrolling the highway saw the plane go down along the Chicopee-Ludlow line by the traffic cloverleaf. The plane skittered across the highway and struck high tension lines in the woods near the Donwell family's farm. There it exploded and covered the highway in burning jet fuel and debris. In addition to lighting the Donwell woods on fire, the KC-135 lit up the night sky throughout the Pioneer Valley with a fierce red glow.[21]

Similar incidents were not uncommon. In November, 1958 as Americans prepped for Thanksgiving, a B-47 bomber and a Stratojet tanker crashed at Loring, both in the same week. The twin crashes claimed eight lives in total.[22] In August, 1959, the eight man crew of a B-52 out of Westover were more lucky, making the first parachute jumps of their lives shortly before their aircraft plowed into a swamp in Epping (near the current site of the New Hampshire Motor Speedway). Near Albany, the plane's radar dome had torn off, flooding the cabin with frigid air. Westover was experiencing heavy weather, making it impossible to turn around. Instead the plane was ordered to Goose Bay in Newfoundland, but crashed en route. The aircrew was rescued by local farmers.[23]

From time to time, aircraft strayed into populated areas and posed a real threat to citizens on the ground below. A T-33 trainer en route from Grenier crashed into a housing development on Worcester's Forest Street, in 1957, destroying six homes. Miraculously only the two crew members onboard lost their lives. All six property owners involved subsequently sued the military for $80,000 in damages, in separate suits.[24]

An even more alarming crash, in October, 1959, resulted in the loss an F-102 Delta Dagger fighter jet in the woods between Andersonville and Barton Village, Vermont. Twenty Air Force police officers kept up, "a jittery vigil," through the night, hoping that stray signals from ham radio operators wouldn't set off the plane's full load of live missiles. The aircraft's 28 year old pilot parachuted to safety and then walked to a nearby farm house to call the Air Force for help.[25]

A B-52 aircrew from Westover was court martialed in July, 1959, for dumping unexploded 20 millimeter ammunition into the Swift River near the Quabbin Reservoir to conceal the fact that they hadn't used up the shells during a training exercise. The Quabbin Reservoir itself had served as a bomb range during World War II and into the early years of the Cold War (pilots used dud bombs filled with sand to bomb the remnants of flooding homes, left by the construction of the reservoir in the 1930s), but by the late '50s, the area was definitively off limits for weapons.[26]

Although the Westover crew was court-martialed, the military had a stunningly relaxed policy to lose munitions throughout the '50s. In one startling example, the Benoit family of Charlton, Massachusetts discovered five 250-pound bombs on the margin of their farm on February 15, 1955. The chief of police collected the weapons and handed them over to the Army, who sent a truck down from Fort Devens to pick them up. Evidently, a plane had dropped the bombs to conserve fuel.[27]

Other troubles were a bit more mundane, but still obnoxious. Sonic booms disturbed people at all hours of the day as the first generation of supersonic aircraft shrieked through the skies overhead. At eight o'clock in the evening, on September 11, 1958, an F-104 Starfighter streaked between Springfield and Boston—covering the distance in six minutes. Police switchboards were soon "deluged" with calls from concerned residents between Westover and Boston, as far north as Saugus and as far south as Connecticut. In a typical move, Otis and Hanscom both denied that the jets involved were from their bases. Westover finally fessed up and admitted to the flyover. A Westover spokesperson told reporters, "If the Russians try to come over, we have to be ready with planes of this type." In the 21st century, modern fighter aircraft are restricted from flying over supersonically, except over the ocean.[28]

Although sonic booms were incredibly common, they had the potential to cause destruction, generating powerful shockwaves which could unseat cookware or even crack walls. An F-104 from Westover flew to Canada to demonstrate to the potential of supersonic technology to the Canadian government. Although the pilot, Captain G. L. Schulstad, was told not to fly supersonic, he accidentally opened the throttle too wide during the demonstration, breaking all of the windows on one side of Ottawa's Uplands airport. The damage soon climbed over $5,000.[29]

To make up for many inconveniences, the Air Force was always keen to stage a good show to impress on the public the importance of supersonic technology. A squadron of Thunderbirds F-100 Super Sabres arrived at Otis for a 'Festival of Fun' in August, 1959, to perform jaw dropping rolls, loops, and U-turns—all particularly impressive given the difficulty of flying the stubborn F-100 air frame.[30]

The Air Force was also eager to impress foreign leaders (although not in the same way as the dramatic and embarrassing demonstration before the Canadian government). Mohammed Reza Pahlavi, the Shah of Iran, arrived at Westover for a tour on June 29, 1958, stopping first at Logan aboard his C-97. Pahlavi was a talented pilot in his own right and was given control of a B-52 briefly on a flight between Westover and Minneapolis (he was treated to 12 days of SAC base tours all across the country). At the time, Iran was a staunch US ally, viewed as a vital bulwark against communism in the Middle East.[31]

The arrival of the Shah in New England was certainly an unusual epoch in the saga of Cold War New England, but pairing off with Persian monarchs comes the tale of the region's own 'red menace.' As it is today, New England in the 1950s was among the most liberal-minded places in the US. Throughout the 1910s, anarchists had exploited worker sympathies at the region's numerous mills, even turning violent with periodic bombings around the time of World War I. During the 1920s and '30s, Southeastern Massachusetts was one of only a few places in the country to court communists and even socialists for local office.

By the end of World War II, the communists were on the wane, but owing to the US alliance with the Soviet Union were still allowed to operate freely. That began to change quickly with the advent of the Cold War, the rise of McCarthyism and communist 'witch hunts' that raged throughout the '50s.

The tide turned against New England communists quickly, and ferociously, leaving lasting effects. In fact, the Massachusetts General Laws still list the Communist Party as a forbidden, subversive organization to this day. For New Englanders involved in socialist and communist activity, life quickly became a living hell.

Take for instance Joe Chase, a railroad worker and communist from Brookline, hauled before the House Un-American Activities Committee in 1958. Along with his brother, Homer, Joe had been raised in a communist household in the 1910s and '20s. His mother, Elba Korb Chase, was the leader of the New Hampshire Communist Party in the 1930s.

In front of HUAC, Joe was asked about his involvement in Lodge 361 of the Brotherhood of Railroad Trainmen. The government's Lead Counsel grilled Joe—an interrogation captured in extensive transcripts. "Mr. Chairman, so that the record may be clear, I should like to now make it abundantly clear for purposes which may arise after this committee leaves Boston. One of the reasons why I am seeking to elicit from this witness a confirmation of his employment is that we understand that he is, or has been in the recent past, employed on a vital industry, namely, a railroad, the New York, New Haven & Hartford Railroad. We would like that confirmation from this witness so that if he does give us the confirmation, we, having that confirmation from this witness so that if he does give us on the confirmation, we, having testimony that he has been identified as a member of the Communist conspiracy, would like to take that information back to Washington to be appraised in connection with other information to perhaps preclude from vital communications, transportation, or other defense facilities, persons who are members of the conspiratorial operations directed from Moscow."

While it was true that Moscow had done everything in its power to control and recruit communist sympathizers around the globe to do its bidding, even the harshest New England Stalinists proved to be little more than outspoken labor advocates. In response to

Arens' framing, Joe Chase replied, "Hold on—Could you shorten that up a little bit? You mean you would like to get my job. Is that you mean."[32]

At the close of the '50s, people prepared reluctantly for the 1960s. On one hand, the '50s had been an incredible decade, replete with jobs, thousands of cheerful new homes, and innovations like television that lightened the mood after the long years of the Great Depression and World War II. On the other hand, the communist menace seemed to be growing and with each passing year the Soviets built better bombs, better planes and better missiles poised to strike down the cheerful American lifestyle of the time. The Cold War was getting colder—a chill was descending everywhere.

# Civil Defense: Guarding the Home Front

Instilling a belief in American-style democracy was part of the Civil Defense plan for the home front in the '50s. These students seem a bit skeptical!

Civil Defense posters exhorted civilians to prepare for the Soviet threat in all its forms, including the possibility of gas attacks on cities and efforts to intentionally mislead the public and incite fear through false news stories.

# Nike Air Defense: Defending a Fortress Without a Roof

Nike Ajax missile battery in Nahant, Massachusetts. Nuclear and non-nuclear Ajax missiles surrounded Boston and Providence to shoot down Soviet bombers.

Missile command post at Fort Heath in Winthrop, Massachusetts in the early '60s.

Missile control area atop World War II casements in Hull-Weymouth area at Fort Duvall.

## Hingham: The Naval Arsenal

The following photos were taken by photographers working for the federal government showing many of the remaining buildings on the base circa 1970, almost 10 years after the Hingham Naval Ammunition Depot closed for good.

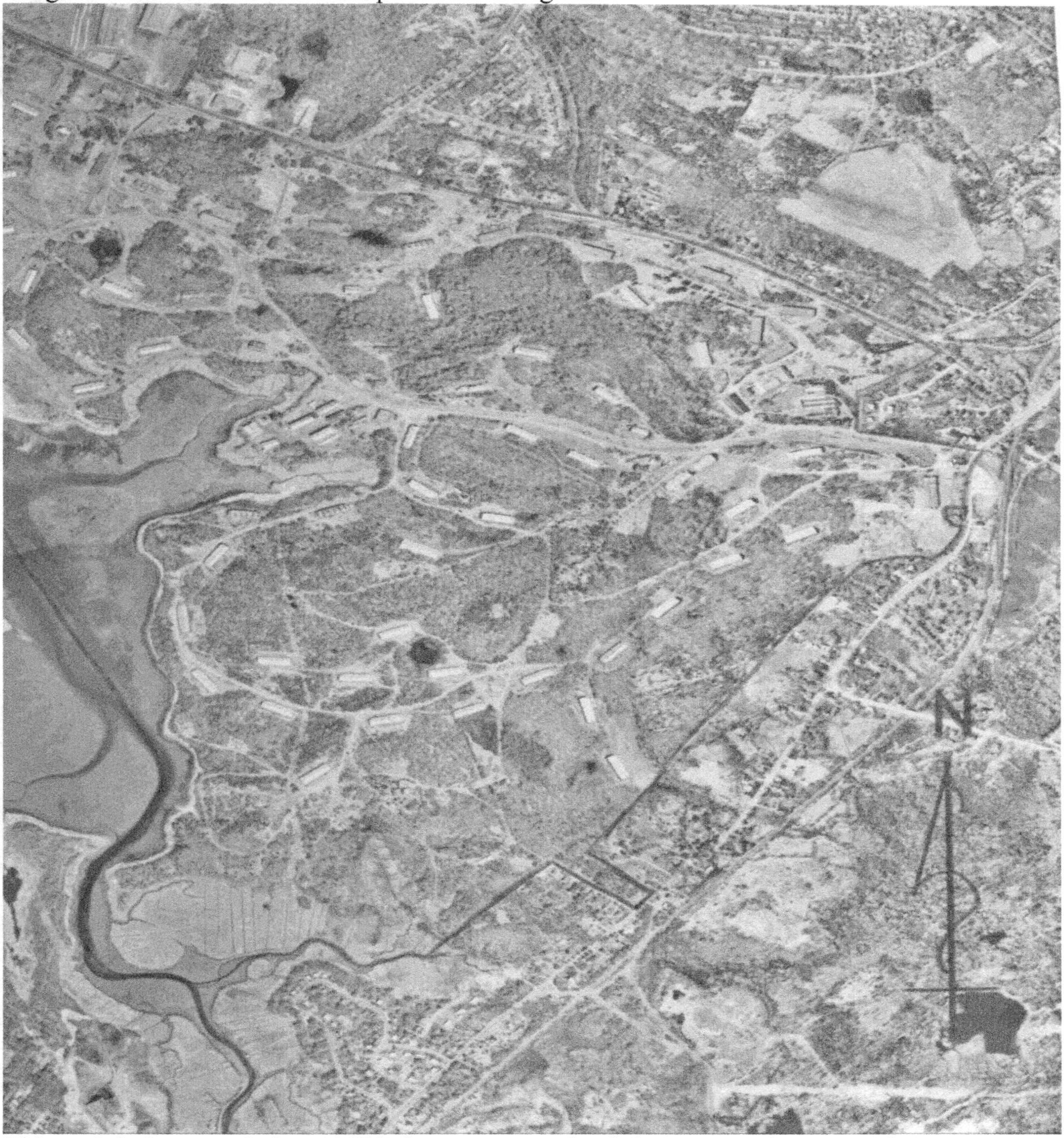


This aerial photograph from 1970 shows a network of above ground ammunition magazines connected by roads and a short rail-line.

A typical underground magazine common at the base.

One of many above ground ammunition magazines, with overgrown railroad tracks in the foreground.

A Mass. Fish & Wildlife worker mows brambles in the late ‘60s or early ‘70s at the Naval Annex in Cohasset.

The commanding officer’s once splendid house sits vacant in 1970.

This wood-frame garage looks like it needs a good coat of paint—or a wrecking ball—in this assessor's photo from 1970.

# Ethan Allen & Plattsburgh: Green Mountain Outposts

Alert hangars for fighters, Ethan Allen AFB. (Courtesy Army Corps of Engineers)

Vermont Air National Guard F-51H Mustangs from the 134th Fighter Squadron in Burlington, 1952, at the height of the Korean War.

134th Fighter-Interceptor Squadron, F-106 supersonic interceptors staged in Burlington in the late '70s.

F-4 Phantoms from the 134th Tactical Fighter Squadron stage at Plattsburgh AFB on June 16, 1984.

## Chapter II: The Scary '60s

When President Kennedy, a Boston native, was elected president in 1960, New England Catholics had good reason to celebrate. The first Catholic president was charming, charismatic and younger than any other recent president. Hopes ran high that Kennedy's administration would be as good as Eisenhower's. However, the years of the Kennedy administration proved more unstable than many people expected.

The 1960s started without a bang in New England. Groton added another feather to their cap, when Don Walsh, a Navy lieutenant, rode to the deepest point in the ocean—the Challenger Deep of the Mariana's Trench—with Swiss scientist Jacques Piccard, aboard the bathyscaph that the Piccard family had constructed. The historic descent involved plunging over seven miles below the surface of the ocean, onboard a steel sphere suspended beneath a large tank of gasoline, capable of resisting the crushing of the deep ocean and providing buoyancy. Ironically, Walsh was technically listed for shore duty, which negated his hazard pay for submarine service. As a 1956 graduate of the Submarine School, the Connecticut base was quick to claim him as their own.[33]

In March, 'weekend warrior' pilots fired up the engines on their Grumman S2F Tracker maritime patrol planes at South Weymouth NAS to fly to Norfolk, Virginia for training. Some of the best performing pilots from the Virginia exercises were picked to take part in a major anti-submarine warfare exercise at Maine's Brunswick NAS during the first two weeks of May. Eight hundred men gathered at the air field for mock missions ranging far out to sea, part of a painstaking rehearsal for actual actions against enemy submarines. Massachusetts-based S2F Trackers, under the command of an officer from West Acton, flew into Brunswick together with longer range P2V Neptune planes from other units.[34]

In addition to training and anti-submarine activities, South Weymouth's Naval Air Development Unit spent the summer of 1960 testing its 'skyhook' device in Alaska (the test unit on the ground in Alaska was headed by a man from Fall River). Designed by Robert Fulton, from Newtown, Connecticut, the skyhook used a helium balloon to hoist a nylon cord and a large hook that would ascend 300 feet, allowing a low-flying plane to snag it and lift the person to safety (a version of the skyhook appeared the 1965 James Bond movie *Thunderball*). The group in Alaska successfully escaped from small islands and even the deck of a ship using the skyhook.[35]

At the end of the year, in December, the small city of Keene, New Hampshire banded together to support orphans in South Korea. Keene High School held back-to-back benefit dances to raise money, while churches collected food, and elderly women in a retirement home knitted mittens. Together with a helicopter from Ft. Devens, a local laundromat helped to haul the supplies to Grenier for a flight aboard a C-97 Stratofreighter. Keene's booming shoe factories chipped in some new pairs for the children, together with local toymakers, who were horrified to discover that South Korean children often played with live hand-grenades left over from the Korean War. A piano, phonograph, and 20 school desks were also part of the load.[36]

1961 started off cold and uncertain. At the end of January, 250 reservists flew C-119 Flying Boxcars out of Hanscom to bases in France, so that they would be able to ferry troops to West Germany in the event of trouble in East Germany.[37] The sonic booms that had terrorized New Englanders in the '50s kept up unabated. On February 19, an F-106 Delta Dart streaking over Boston at 40,000 feet set off a sonic boom heard as far north as Lynn and as far south as Norwood. The boom shook windows, roused dozens of babies at Massachusetts General Hospital, and clogged emergency lines throughout the city with people phoning in about an explosion.[38]

Although the nationwide campaign to exterminate the polio virus had made tremendous gains since the introduction of Jonas Salk's famous vaccine in 1955, polio prevention still had a very long way to go in developing countries. In spite of Cold War tensions, the US Navy distributed doses of polio serum to Cuban officials at the gate of Guantanamo Bay, at the start of March, 1961. A group of Naval Reservists from South Weymouth flew down to Guantanamo to witness the moment, and helped the Navy to hand over the serum.[39]

Perhaps one of the strangest stories of the Cold War began to materialize starting in 1961 and continuing through 1963. Project West Ford, also known as Project Needles, found the US military attempting to create a massive artificial ionosphere above the Earth's atmosphere to boost military communications. The Pentagon was afraid of inconsistent communications with telecommunications running through undersea cables that the Soviets could easily cut, or bouncing off of the Earth's unpredictable natural ionosphere in the upper atmosphere.

Dubbed West Ford, after Westford, Massachusetts where the project's receiving parabolic antennae would be located, the crazy scheme involved launching 480 million copper dipole antennas—each about the length of a postage stamp—into orbit. The project was first dreamed up in 1958 by Walter Morrow from MIT's Lincoln Labs and the dimensions of the pins with an 8 GHz signal in mind. Following an unsuccessful first launch in 1961, where the pins failed to disperse, subsequent launches distributed enough pins that the system was able to send signals by May 9, 1963. Although the project was shelved soon after, it lives on as one of the biggest contributors to space junk with clumps of pins still orbiting the Earth today.

In the spring of 1961, steel workers at the Fore River Shipyard in Quincy cheered along with the Navy brass. The Bethlehem Steel Co.'s shipyard heralded the upcoming launch of the nation's first nuclear guided missile frigate—one of only three nuclear surface ships in the Navy at the time. The launch was attended by the great-great granddaughter of American naval hero, Commodore William Bainbridge, famed for his exploits in the War of 1812, for whom the ship was named. The 550-foot frigate was a far cry from Bainbridge's original vessel, armed with Terrier missiles and advanced anti-submarine weapons instead of cannons. While *USS Bainbridge* made its way down the slip and into the Fore River, another, much large ship took shape in a dry dock close by.[40]

*USS Long Beach*, the Navy's first and only nuclear cruiser (and the only ship of the *Long Beach*-class) was laid down in Quincy starting on December 2, 1957, and was commissioned in September, 1961.

As East German construction workers began to assemble the Berlin Wall in August, 1961, service members in New England patrolled and trained around the clock. With dozens of flights a day going between regional air bases, accidents were bound to happen. On August 12, a KC-135 tanker returning to Westover ran into danger when Major Howard Lippincott, the plane's pilot, discovered that the plane's nose wheel was not descending as he prepared to land. Running on fumes, Lippincott radioed to fly to Griffiss AFB in Rome, New York. A B-52 bomber, bound for Westover and two B-47s from Pease AFB diverted to escort the stricken tanker to Rome. Reversing the usual arrangement, the pilot of the B-52, extended a hose to the KC-135 and by flying above it, used gravity to pump some fuel into the tanker's nearly empty tanks.[41]

The same day, hundreds of New England National Guard troops boarded C-119 Flying Boxcar transports at Hanscom and Grenier to fly to bases in South Carolina. Once they arrived, the soldiers took part in one of the largest practice airdrops since World War II. Six thousand paratroopers, the majority from the 101st Airborne Division took the plunge out of the Flying Boxcars. The first man to jump was 51 year old Major General Charles Rich. "We hit the drop zone on the nose. We think the Army will be very happy with the drop," Brockton native, Dona LaCouture Jr. told reporters.[42]

In September, Groton launched its latest submarine, the *USS Thresher*. Hailed as the world's most advanced sub, Thresher's shakedown was described as having gone "exceedingly well." Eight of the vessel's nine officers were residents of Connecticut, along with 80 enlisted sailors. Commander Axene, who had previously served aboard the *Nautilus* was asked to describe the difference between the two nuclear submarines. "[It's] like trying to choose between your wife and your mother."[43]

Later in the month, the Air Force premiered its own breakthrough, connected with Hanscom. A 134-foot tall radome was erected on the side of a hill in Tyngsboro—largest radome anywhere in the world. The 120 foot wide antenna dish and accompanying electronics were supplied by MIT's Lincoln Laboratory, with the expectation that they would go 'on-line' in 1962. The purpose of the radome would be to join the Millstone Hill satellite tracking station in Westford, to develop increasingly advanced technology to track and communicate with first generation satellites.[44]

A major air exercise took place in the skies over Boston in October. An Otis pilot scored the first air-to-air 'kill' of Operation Skyshield II over Massachusetts Bay. For the exercise, all commercial flights across the US were cancelled—2900 in total—and aircraft scrambled from every major New England air base to respond to the mock 'Red' threat. Commercial flights were cancelled as a result of widespread NATO signals jamming, which could have disrupted the LORAN radio signals that planes used to navigate (a predecessor of GPS that beamed out different frequencies for different latitudes).

The 'attack' started with emergency alerts from the DEW (Distant Early Warning) radar line in the Arctic. Soon thereafter, 25 F-101 Voodoo fighters streaked out of Otis to defend Boston from 250 B-52s and B-47s, accompanied by 15 British Avro Vulcan bombers. On the ground in Boston, 150,000 people gathered at Logan to see the aircraft coming and going from the offshore exercise. In a real conflict, there would have been few spectators left alive—the simulation included 'missile strikes' on the major airports throughout New England.[45]

At Westover, 1961 witnessed the introduction of a first generation computerized meteorology system. System 433L would automatically record weather data and make it available for fighters and missiles. First dreamed up in 1958 as a nationwide, all services collaboration with Federal Aviation Administration and the Weather Bureau, the Air Force was left as the last one standing with the project. "There was delay after delay in our discussions and the Air Force felt it could wait no longer," explained one officer, going on to add that the Air Force was proud of the result, which could transmit 2,000 words per minute.[46]

A few days before Christmas, 1961, Roscoe Ammon, a Manchester industrialist died at New England Deaconess Hospital in Boston, at the age of only 43. Ammon had donated half a million dollars to the city to allow them to conduct civilian redevelopment at Grenier, after the base's partial closure in 1960 (the base remained officially on the Air Force rolls in a limited capacity until 1966).

As the owner of Marion Electrical instrument company, Ammon churned out 30 different patented devices, including a major navigational component used in the Mercury space program and the X-15 space plane. In honor of the late inventor, Manchester agreed to name Grenier's new control tower after him.[47]

While the rest of the nation celebrated Christmas, SAC remained on high-alert. Because the Soviet Union adhered to state atheism, they made no official efforts to honor holidays such as Christmas. Throughout the '60s, '70s, and '80s, Christmas Eve and Christmas Day were viewed as the most likely times for a Soviet surprise attack. SAC regulations permitted only 20 percent of its personnel to be off-base at any given time. That meant that on Christmas, 80 percent of its personnel remained at their stations at Westover, Pease, Dow and Loring. Even on Christmas, crews were told to expect training alerts—sudden orders to scramble out over the ocean in preparation for a nuclear assault on Eastern Europe.[48]

1962's Armed Forces day was one of the most exciting displays since the shows of the '50s. Millions of New Englanders turned out for tours and airshows, including a public day at Groton and a glimpse inside the new guided missile cruisers in Boston. At the Army's Quartermaster Research and Engineering labs in Natick, the largest solar furnace in the hemisphere was on display, as well as the latest equipment rolled out for the military. The only disruption to the day's events was at Hanscom, where an Air Force engineer suffered a sudden heart attack and died. On the runway, five C-119s taking off together nearly collided, but fortunately no more lives were lost.[49]

The excitement of Armed Forces day was dimmed somewhat by the revelation, 11 days later, that three fifths of the state population would die in the event of a nuclear war.

A group of physicians in Boston conducted the first comprehensive study of the possible effects of a nuclear war in the state. The results were certainly not encouraging. Out of a statewide population of 4,691,000, an astonishing 1,300,000 would die instantaneously in Boston, Worcester, Quincy, Springfield, New Bedford and in the communities right around major bases. Two million eight hundred thousand more would be seriously injured from burns, falling debris, and exposure to radiation, leaving only 965,000 people from the most rural parts of the state to care for them. According to the committee, only 2400 of the state's 9440 physicians would survive the attack, and fewer than 10,000 of the state's 65,000 hospital beds would still be available.

Many of the state's psychiatric hospitals were located in more rural areas, but what would happen if those patients were turned loose to make way for the injured? To make matters worse, the study concluded that the state's medical stockpiles (if they survived) would only be enough for 15 percent of the injured. Even more frighteningly, the study assumed a *light* attack, with two 10-megaton bombs falling on Boston, one each on Hanscom, Westover, Otis, and smaller bombs falling on Fitchburg, Fall River and Holyoke.[50]

In July, the New Hampshire government set aside former military barracks at Grenier to house African-American refugees from the Deep South, attempting to escape the violence surrounding the civil rights movement. Seven refugees arrived at the base, but to the surprise of the New Hampshire government, quickly moved on, finding jobs at Rye Beach hotels, Nashua shoe factories, a trucking company, and a restaurant in Keene.[51]

August, 1962, witnessed a deadly tanker crash in Concord, Massachusetts, close to Hanscom. A KC-135 coming in to land undershot the runway and fell into neighboring woods. None of the crew survived. Fortunately, it seemed as if the tanker had offloaded most of its fuel before landing, preventing a much more serious catastrophe. Nonetheless, it took firefighters from five different towns to subdue the blaze.[52]

The tanker crash didn't help to assuage the fears of local fire chiefs, who had gathered at Hanscom in June for what turned out to be a horrifying display. Six hundred chiefs had gathered to watch a presentation by three Air Force specialists, who set off mixtures of liquid oxygen and motor oil, sending small mushroom clouds of smoke soaring into the air. Even more frightening, the specialists set off TEA (triethylaluminum) a solid igniter for rocket engines with no known way to extinguish it. Adding water to TEA, it turned out, would cause a much more severe explosion. New fuels for missiles and rockets were pouring out of Hanscom and hitting the roads every day of the week, bound for Cape Canaveral. For small local fire departments, the new traffic raised the specter of disaster.[53]

August was also marked by a military milk rigging scandal involving the Boston Navy Shipyard, Portsmouth, Hanscom, and Otis. The Justice Department filed criminal charges against major New England milk producers for violation of anti-trust laws. They alleged that between 1957 and 1959, United Farmers of New England, H.P. Hood, and the

Whiting Milk Co. had deliberately manipulated the price of milk and thereby 'defrauded' the government in large sales to military bases.[54]

In 1961, the US backed an unsuccessful invasion of Cuba at the Bay of Pigs led by Cuban exiles. A year later, Cuba was back in the news again. This time, the island nation was at the center of a nuclear crisis as the Soviets set up missile launch sites while importing nuclear warheads and troops. The Soviets had already built secret bases in East Germany to counter US missiles in Turkey and elsewhere in Europe. As the US demanded a withdrawal of Soviet missiles and "quarantined" Cuba with warships, the two nations came the closest to nuclear war that they had ever come before. When the US and USSR negotiated an end to the perilous Cuban missile crisis, both nations were within days—some historians argue, hours—of nuclear war.

At Westover, SAC's major installation in Chicopee, the first evidence of something out of the ordinary came when one pilot was ordered by his commanding officer to fly film to Rochester, New York, at a moment's notice for analysis. The pilot could not avoid overhearing his commanding officer talking about Cuba.

In preparation for a full-scale nuclear assault on Westover, SAC moved its command and control operations underground—literally. The Post Attack Command and Control System Facility (PACCS), better known as "the Notch," was a hardened underground bunker, designed to allow SAC's 8th Air Force to continue operations in the aftermath of an attack. The facility, built in 1957 in the side of Bare Mountain in Hadley, Massachusetts stayed operational until 1970, but saw its most significant use during the

Cuban Missile Crisis, when it became a hive of activity. The Notch was one of the earliest major post-attack facilities in the US, equipped with a microwave transmitter to communicate with Westover.

The Notch was only a small part of the national PACCS system; a system run by the National Command Authority (a group of national defense leaders including the President and the Secretary of Defense). Other bunkers and command posts were set up at the other major SAC bases across the country to communicate with SAC headquarters in Nebraska, and circling communications aircraft that would stay airborne during a crisis.

At Logan, 12 B-47s were distributed from Pease to make it harder for the Soviets to target all the region's bombers at once. SAC cancelled leave for all of its New England personnel, and ordered its members to position themselves near telephones to prepare for an alert. Under tight security, the Navy sailed its destroyers and submarines out of Boston and Newport to reinforce the Cuban quarantine. Two Massachusetts Nike missile batteries were placed on alert, with weapons pointed skyward. Other nuclear-equipped Nike Hercules posts were told to standby.[55]

Through a combination of luck and diplomacy, the war never came to pass. Instead, the only battle that was won was the fight with the insects at Dow. There, long-term concerns about bird strikes were finally put aside after careful lawn mowing near runways and the application of copious amounts of pesticides eliminated the insects that birds liked feeding on![56]

The same year as the Cuban missile crisis, the Air Force completed work on an Atlas-F ICBM missile field that spanned the Champlain Valley between upstate New York and Vermont. The only missile field ever built east of the Mississippi River included 12 missile silos—two in Vermont, and 10 in New York—loosely arranged around Plattsburgh AFB. The silos opened for operation in December, 1962.

The Cuban missile crisis was a terrifying experience for both countries and led to new treaties to prevent above ground nuclear testing. In spite of how narrowly war was averted both countries remained confrontational.

In response to the recent Cuban missile crisis and the now ever-present threat of nuclear war, the New England Committee for Nonviolent Action decided that it would picket Electric Boat at the start of January, 1963. A 24 year old veteran protestor who had recently spent a month in a California jail for a similar protest told reporters of his plan to sneak into the plant and hand out leaflets to workers. "I hope by doing this to cause EB workers and visitors to seriously consider what they're doing, and therefore maybe they will be as appalled by it as I am."[57]

On January 24, 1963, a B-52 out of Westover lost its vertical stabilizer due to turbulence and crashed into the side of Elephant Mountain in Piscataquis County, Maine. Out of a crew of nine, only the pilot and navigator survived. Below zero temperatures complicated rescue efforts. A bulldozer operator near Mt. Katahdin tipped off the Air Force, telling them that he had seen a towering column of black smoke rising from the woods 10-miles away. A tanker from Dow and a Fish and Game plane deployed to take a look, but didn't catch a glimpse of the bomber, due to thick tree cover, five foot snow drifts and temperatures that were already plunging to negative 20 degrees. The following day, the nearby Scott Paper Company dispatched its workers armed with snowshoes, plows and dogsleds to find the bomber, which they did by mid-morning. The survivors were airlifted to a hospital by helicopter. Incredibly, given the remoteness of the crash site, the remnants of the bomber were left in place, and remain there to the present day as a memorial to the deceased crewmembers.[58]

1963 was marked by several other high profile accidents. In March, a mechanical failure ejected a B-52 navigator over Springfield, although his parachute saved his life. However, a true disaster took place on April 10, 1963, in the Gulf of Maine.

On April 9, *USS Thresher* sailed out of Portsmouth for dive trials after her overhaul, accompanied by the rescue ship *Skylark*. The *Thresher* dived beneath *Skylark*, proceeding downward in a slow spiral, stopping every 100 feet to signal to surface. When the submarine reached its test depth, *Skylark* received a garbled communication "minor difficulties, have positive up-angle, attempting to blow." When *Skylark* crew stopped receiving communications, they quickly realized that *Thresher* had sunk, with all 129 hands. Fifteen Navy vessels were called in to participate in the search for the sunken submarine.

An extensive Navy investigation followed, in which the *Trieste* arrived in Boston aboard a barge and deployed twice, first in June and then in August, 1963, recovering photos and pieces of the *Thresher*. The Navy concluded that the submarine had suffered a sudden failure of its salt-water piping system, when a silver brazing fractured.

In the years since the Thresher disaster—one of only two American nuclear submarines ever to have sunk—the Navy has conducted extensive studies of sea floor sediment to check for radioactive leaks from the vessel's reactors, which now lies at 8,400 feet deep with the rest of the wreckage.

Many years after the accident, in 1985, Woods Hole Oceanographic Institution researcher Robert Ballard was commissioned by the Navy for a secret plan to photograph the *Thresher* and collect photos from the wreckage of the *USS Scorpion*, a submarine which sank near the Azores in 1968. The cover for the project was Ballard's mission to find the *Titanic*—which he did, with the use of the same technology used to find the two submarines.

1963 also brought strange and alarming changes off of New England's coasts. As early as 1963, Soviet fishing vessels had started to appear in large numbers on Newfoundland's Grand Bank and the George's Bank, New England's slightly smaller fishery. In August, 1963, Soviet trawlers tied up in Halifax, Nova Scotia harbor. In the words of the *Boston Globe*, "For the east coast port, steeped in the traditions of the sea in war and peace, is becoming a home away from home for hundreds of Russian sailors and their ships."

Halifax truly was becoming a 'home away from home,' for the Russians. The first Soviet ships had arrived in 1959, in search of cod, herring and redfish to feed a burgeoning population in the USSR. Almost immediately, the Soviets assembled their trawlers into fishing fleets, organized around massive factory ships—floating canneries that took deliveries from orbiting trawlers. By 1963, 30 to 50 Soviet ships were operating out of Halifax on both the Georges and Grand Bank. The fleets were becoming even more organized, with water tankers to supply trawlers and canning lines and 'draggers' sent to scour the sea floor for bottom feeders.

Fishing season ran between February and November. At peak activity in June, a local service agent reported at least one ship mooring each day in Halifax harbor to take on supplies, although ships typically arrived in groups of two or three. Canadian store owners welcomed Soviet rubles (converted over to dollars in port) and the polite Russian sailors that were eager to spend them. In fact, each Soviet ship often spent as much as a thousand dollars on supplies during a visit to port. The Russians were known for their soccer games on the wharfs, large alcohol purchases, and an eagerness to stock up on nylon women's stockings and fresh vegetables (nylons were a consumer product that the USSR's state-run factories could never seem to produce) from the US.

To the surprise of the Canadian officials that dealt with the Russians most often, none of the Russian sailors tried to sleep away to start a new life in the West, although the Canadians did notice that Russian sailors traveled everywhere in groups of three. Port officials accepted the situation blithely. The Russians, it turned out, were fastidious when it came to paying their customs fees. It seems that the Soviet fleet's 'three-legged stool' approach blocked most sailors from skipping out on their crews, but for some sailors, the

Soviet ships offered incentives to stay. Onboard the largest ships, such as the *Oka*, the largest trawler out of Kalinigrad and the water tanker *Vyru*, out of Tallinn, Estonia, many crewmembers had private rooms completed with desks and lamps.[59]

While the Canadians were tolerant of the Soviet presence, their American neighbors were a bit more hesitant. From the moment it arrived, the Soviet fishing fleet was regarded with a large measure of suspicion, and suspicions were well founded. The American aircraft carrier, *USS Intrepid*, was tailed by a Soviet trawler, "bristling with radar and other sophisticated electronic gear."[60] Many 'trawlers' were equipped with advanced radio receivers, radar and other electronics, which the Soviets insisted were exclusively for the purpose of catching fish. Although some of the electronics undoubtedly were for tracking schools of fish, their true purpose was to record US and Canadian signals intelligence and track the coming and going of submarines and surface warships.

As early as the fall of 1963, fishing groups in Gloucester and New Bedford demanded that the US territorial limit be extended farther out to sea, to protect fish stocks. Soviet trawlers used finer net mesh than their American counterparts, leading to fears about declining fish stocks—especially ruthlessly harvested flounder, a bottom fish. Newly elected Senator Ted Kennedy was quick to endorse the idea, along with Massachusetts Republican, Representative F. Bradford Morse, who told reporters that the presence of Soviet ships, "indicates a Soviet determination to spy on NATO military maneuvers." As if to affirm Morse's belief, a Soviet trawler and tender anchored two miles off the shore of Truro, Massachusetts, for 'rudder repair' and then made a quick get away into thick fog and international waters. Given the proximity of Truro Air Force Station, it seems that the Soviets might have been collecting intelligence.

The Soviets were as numerous, or in some cases even more numerous than American ships on George's Bank, but they were only part of the foreign presence. Coast Guard observers tallied at least 169 foreign ships by the end of 1963, with estimates suggesting that as many as 300 were present. Ships came primarily from Europe, to feed blooming populations hungry for canned seafood. Some trawlers were from Britain, France, Spain and West Germany, but others sailed from East Germany and the People's Republic of Poland.[61]

The House Armed Services Committee soon took up the concerns of New England fishing crews, as fears grew that the Soviets might land spies and saboteurs or interfere with anti-submarine countermeasures.[62]

In September, 1963, Manuel F. Lewis, secretary of the Gloucester Fisheries Commission charged that the Soviets were deliberately destroying New England fisheries, as an effort to harm the US economy. Brockton Congressman Hastings Keith agreed. "We are engaged in economic warfare with the Communist Bloc, and we must realize that Georges Bank and Nantucket Shoals are battle grounds in this war."[63] Lewis subsequently appeared before a Senate committee to suggest impounding Soviet ships that strayed into US fisheries and charging a hefty, $10,000 fine for their release.

The start of 1964 came with a bang, although not quite as large of a bang as might have been. On January 18, a B-52 from Westover crashed into Big Savage Mountain, part of the Appalachian Mountains in Cumberland, Maryland with nuclear weapons onboard. Rescue workers and explosive ordnance demolition teams raced to the site, where they were able to report, "no danger of nuclear explosion." Although the danger of an explosion might have been minimal, it was still a harrowing incident, which could have contaminated western Maryland with hazardous, and potentially event fatal, concentrations of radiation.[64]

In February, NASA announced a $50 million electronics research center in Cambridge, which was expected to bring in $250 million for the local economy. The site was selected out for its proximity to Harvard, MIT, and BU as a location for designing the electronics needed to launch the space race in earnest. Other communities that had been campaigning for the NASA facility were sorely disappointed. Iowa's congressional delegation retreated crestfallen, along with New York's—the New York delegation had supported the Massachusetts bid, so long as the facility was built at Hanscom, which would allow Hanscom's research and development work to transfer to Griffiss AFB in upstate New York.[65]

To add insult to injury, the Navy commissioned a new Polaris sub facility at Melville, Rhode Island, which was destined to attract 3,000 service members and their families to Narragansett Bay. Hanscom premiered a $30 million radar center at the end of March, capable of tracking missiles, satellites, and the first space debris (although to the satisfaction of other delegations, this facility was actually assembled at Eglin AFB in Florida).[66]

It seemed that the only opponents of the growth were a handful of pacifists, easily brushed aside by the government. In May, a federal marshal condemned the rowboat *World Citizen* in Groton. The 12-foot long, $40 boat, with its twin oars and three life jackets had served as the sum total of equipment available to pacifists in 1961, protesting the Polaris missile subs. The two pacifists involved had been arrested and convicted of trespassing, but the boat had languished in a dry dock for three years as a result of a court battle between the Navy and the boat's owner in New York.[67]

In July, a fighter pilot from Newburgh, New York managed to save over 150 lives in Western Massachusetts, when his fighter plane went out of control. Thirty-three year old Captain John Paris saw that his plane was veering toward a public beach. He steered his plane clear of the beach and clear of the Barnes field and ejected at the last second at 1200 feet. Paris fell to his death on the shore of Hampden Pond, while his jet crashed harmlessly in an open field.[68]

Offshore, the Soviet fishing fleet continued to inspire anxiety. At the end of September, NATO officials announced their belief that the ships represented a small fraction of a massive intelligence fleet extending across the Atlantic Ocean to Iceland, Britain and Norway. Coast Guard observers began to pick out "AGIs," intelligence-gathering auxiliary ships masquerading as trawlers. The tip off for an AGI was its speed. Although it might look like a trawler, it could sail at the same speed as a destroyer at full thrust. During a war game in Canada's Bay of Fundy, AGIs raced in to follow NATO anti-submarine warfare ships and carriers. AGIs also tailed the MERCONVEX fleet, a chain of

22 ships that rehearsed the largest freighter convoy crossing of the Atlantic since World War II. One British officer escorting the column quipped, "These chaps aren't fishing...I've never seen a fishing craft so clean and tidy after weeks at sea. Now they are operating independently all over the ocean to watch our exercises...No fishing boat ever went that fast [25-knots]."[69] Soviet spies taped radio conversations and even propeller noise to game the rotations per minute (RPM) of NATO ships, a key 'signature' feature for identifying ships by sound. Although NATO was silent on the role of other fleets, East German and Polish intelligence may have staged similar deception to feed information back to their Soviet masters.

As 1964 drew to a close, the period of growth at New England bases gave way as closures ramped up. Secretary of Defense Robert McNamara ordered the decommissioning of 95 bases around the country, including Dow, Portsmouth Naval Shipyard, Springfield Armory and many other locations across the country.[70] McNamara agreed to a delayed closing for Portsmouth, over the course of 10 years. The base employed 7,274 workers in 1964. In fact, McNamara promised *increased* construction and repair work at Portsmouth through 1965, to give families enough money to ease into the closing.[71]

1964 closed on an unhappy note in Maine, where protestors gathered at Dow in early December to attempt to stop the closure of the base. "If it is the best national interest and will result in national economies, in good conscience we cannot protest it," City

Manager Joseph Coupal told reporters, measuredly. Nonetheless, Coupal asked Maine's Congressional delegation to intercede.[72]

Answering the call, all six New England governors met with Secretary of Defense Robert McNamara. Springfield Mayor Charles Ryan joined the protest, declaring that McNamara's decision to close the Armory was, "a flagrant disregard for the economic life of one section of the country."[73]

Perhaps as a concessionary move, or simply because of technical expertise in the Boston area, the Air Force inked a contract with Perini Corp. of Framingham in February, 1965. Perini Corp. would begin constructing a ballistic missile field, consisting of 50 silos around Great Falls, Montana. Perini was already at work on 550 silos in Missouri, Wyoming, North Dakota and Montana (the missile fields constructed by Perini, in conjunction with other companies remain active to the present day and are likely to house America's nuclear arsenal well into the 21st century).[74]

Atomic Age construction was on the minds of many structural engineers, including some in New England. In addition to the large engineering projects buildings bunkers and bases, some private companies invested in blueprints that would hold up to nuclear attack. Thomas Vaughan, today a Professional Engineer and member of the Society for Industrial Archaeology recalls working on an addition to Paul Revere Insurance in Worcester, built around 1967. The insurance company's new parking garage was designed to double as a fallout shelter in emergencies. With this in mind, roof drains were cast inside the concrete columns so that people hiding inside would not be exposed to fallout washing out from the roof.

Other cutting edge work was happening just up the road in Bedford. John W. Salisbury, a PhD in geology who had passed through the storied halls of Amherst College and Yale before coming to Hanscom headed a team of 10 scientists who conducted research year round and issued the country's only journal of astrobiology and lunar science. Salisbury was the perfect man for the job. He had refused to give up his interest in lunar research when he first arrived at Hanscom in 1957, even when most people chuckled at the thought of a manned landing on the moon. Salisbury had conducted his geology dissertation in the mountains of Georgia, where he had frequently come under fire—literally—from moonshiners who mistook him for a federal agent. He could, and did, persevere, through whatever challenges doubters threw his way.[75]

At the end of 1965, McNamara announced further cuts, including Boston Army Base, which would be shuttered by 1970. The base's operations would move to nearby Fort Devens. Other small military facilities would be consolidated, including the Naval Reserve Training Centers in Woonsocket and Lowell, the Marine Corps Reserve Training Center in Lawrence, the Long Island Fuel Annex, the Casco Bay Naval Station, the Nike missile sites near Caribou, Maine, and the Portsmouth Naval Hospital. Portsmouth's imposing Naval Disciplinary Command prison would close down in 1969, with any naval prisoners transferred to the Army Disciplinary Barracks in Ft. Leavenworth, Kansas.[76]

To add insult to injury, the federal government couldn't even manage to stem the tide of Soviet fishing. After a failed conference at the University of Rhode Island (and not being invited on a single National Fisheries research cruise) New England fishing groups grew increasingly incensed with the Bureau of Commercial Fisheries. Making matters worse, the *SS American Contractor* unloaded blocks of fish in East Boston, in March, labeled 'Product of USSR,' marking the first importation of 'communist cod.'[77]

A modest compromise came in April, 1965, when the Soviets agreed to allow three US inspectors—two from the Fisheries and one from the Coast Guard—to come aboard the trawlers in Halifax in exchange for onshore tours of US Coast Guard bases.[78] American inspectors visited the mother ships *Ivan Fedorov* and *Sviatagor*. As a good will move, the Soviets announced that crews from small American trawlers could use the 20-bed hospital and dentist office onboard each of the mother ships. "Our fishing interests are not mutually exclusive, on the opposite they harmonize with each other," the Soviets announced diplomatically.[79] Inspectors found the Soviets catching undersized cod, but were forced to concede that Americans were doing the same thing.

After the start of reciprocal inspections in 1965, the fisheries dispute died down for a few years, although it remained perpetually on the mind of American fishing crews. In 1966, the *Boston Blue Sheet*, a regional fish market report announced that the Soviets were planning to build the 'Benthos-300,' a 10-person undersea lab on the Georges Bank, as long-term research station to study fish populations (and, presumably, to collect sonar intelligence without the interference from Soviet surface ships). Meanwhile, onshore fishing groups lobbied Congress for a $100 million grant to support struggling fishing communities. The *Boston Globe* weighed in, "While these negotiations are taking place, this fast growing, modern Soviet fishing fleet, with the most advanced equipment, depletes the fishing grounds only a few miles off our coast, and the US fishing fleet, lacking protection, encouragement, and support from our government, withers away."[80]

The Soviets never built their undersea lab, but they did moor a 167 foot trawler (built by an East German yard) at Woods Hole, in 1967, for a joint research cruise with the Oceanographic Institution to Georges Bank and Block Island. While the Americans were eager to meet six of their top Soviet counterparts, the Coast Guard stayed wary of the Soviet ship and sent a cutter to escort it to three mile limit of US waters.[81]

In 1965, following the Kennedy assassination and the start of the Johnson administration, the Gulf of Tonkin incident (an event that later proved to be largely fictitious) led the US to commit to a large scale to a war in Vietnam.

Although the US had already sent advisors to Vietnam, the Gulf of Tonkin incident signaled a move to much larger numbers. Marines were deployed in their thousands to South Vietnam in 1965, as the US launched a three year air war targeting North Vietnam (the nation backing the communist Viet Cong insurgency in South Vietnam), making good on General Curtis LeMay's promise to, "...bomb them back to the stone age". The initially small Marine deployment increased rapidly from a force of only 3,000 to almost a quarter million a year later. Numbers continued to increase throughout the later '60s, until the US had almost half a million "boots on the ground," numbers only matched by US forces in Germany.

The nominal US role in Vietnam was to assist the Army of the Republic of Vietnam (ARVN), South Vietnam's military, but tended to take a leading role in many counter insurgency operations and the bombing campaign against North Vietnam, and rural, communist controlled areas of the South.

1966 marked New England's first taste of full-scale war since Korea 13 years earlier. Base activities switched from the largely defensive operations against the Soviet Union, to preparations for "Nam."

Armed Forces Day was sharply curtailed in May, as units departed for Vietnam. The airshows at Otis and Hanscom were both cancelled, even though they were events that regularly attracted 150,000 people or more. As a consolation, the Air Force opened up its facilities at Westover and Pease for smaller air shows and the Navy announced a more modest open house at the Boston Navy Yard.[82]

Armed Forces Day proved disappointing, but New Englanders were treated to the unusual site of UFOs over Boston and Central Massachusetts on May 20. Hundreds watched as a bright light streaked across the skies, finally dipping out of view over Easthampton and southern Vermont. Dozens of police stations received calls about the strange light. In fact, it wasn't an alien spacecraft, but rather a meteorite that Westover managed to track for a short time.[83]

Although it arrived too late for Armed Forces Day, the aircraft *Wasp* moored in Boston at the South Boston Naval Annex. For a brief 45 minutes, the public was allowed to watch the crew unloading the Gemini spacecraft that *Wasp* had recently retrieved from the ocean. The spacecraft was trucked to South Weymouth to have toxic fuels removed before it was returned to the McDonnell Aircraft plant in Missouri. The spacecraft was left uncovered as it lumbered down the highway to South Weymouth.[84]

At Hanscom, an F-102 shrieked over the field in August, making an incredibly accurate precision "strike." The strike had been called in by a Jeep-mounted mobile radio center, developed by the Electronic Systems Division and dubbed (rather unimaginatively) AN/MRC 107. The new system hinted at the new realities of close air support. General Dynamics announced plans to produce 377 of the units for use in Vietnam.[85]

South Weymouth appeared in the news once again, in September, when workers began tearing down the second largest hangar in the world. The 203 foot high blimp hangar, built during World War II stretched 960 feet long—almost a fifth of a mile. Some saw the demolition plan as a waste of a two million dollar building which amounted to 2.6 million square foot of indoor space.[86]

A week later, a large anti-submarine exercise got underway at Quonset Point, with dozens of Naval aircraft ferrying between the Rhode Island base and Montauk. Meanwhile, Congress considered a plan to call up individual National Guard troops for Vietnam service, within larger units that remained in reserve. Throughout Massachusetts, units were filled to capacity, with people waiting to join the service. In fact, the only vacancies in the state were at the Fall River, Whitinsville, and New Bedford arsenals. Some of the vacancies may have been the result of call-ups.

Hanscom was the center of an emergency airlift of troops to Southeast Asia. One hundred twenty thousand soldiers were airlifted aboard military aircraft, after being stranded by a commercial airline strike.[87]

While the Vietnam War began to get messy, 1967 remained a quiet year for New England bases, with the exception of continued call-ups and deployments. In March and June, Boy Scouts and visiting students from the Brazilian Naval War College toured Groton to get a glimpse of a modern naval base and its resident nuclear subs.[88]

The Navy's public relations gambits in the first half of the year gave way to embarrassment during the second half. At the start of September, a nuclear submarine struck a sloop in Block Island Sound. Fortunately the vessel only suffered a few thousand dollars damage, but Richard Carlson, the vessel's owner suffered a damaged ego. He berated the Navy for bumping into his boat, tipping over a boiling pot of coffee that injured his wife's feet and then leaving the scene without hailing him about assistance. To make matters worse, the Shelter Island native claimed that the submarine had taken no heed of his running lights. The Navy rebutted Carlson's claims with a prepared statement and a brief sound bite for reporters: "[the facts] do not jibe."[89]

A follow-on mess came in November, when the submarine tender *USS Fulton* dumped the accumulated waste from a 1,000 men into the Thames River untreated. The dumping came in strict violation of a Connecticut pollution abatement program that the Navy had casually ignored since 1956.[90]

At other naval installations, things tended to be more even more mundane, but certainly more amusing. Twelve year old Kenneth Baumeister slipped off the operating table at the South Shore Hospital just as his tonsillectomy operation was about to begin and went on the run. Intent on keeping his tonsils, Kenneth's escape led to a search by local police and even sailors from South Weymouth. In spite of his daring evasion, Kenneth's tonsils ultimately lost. When he returned home, Kenneth's parents promised him another date with the surgeon as part of a package deal for seven of the nine Baumeister children to have their tonsils removed.[91]

Bad luck continued for Groton into 1968. In March, a survey concluded that water in nuclear ports was showing increased radioactivity. Navy scientists noted a buildup in the radioactive isotope cobalt-60 in harbor bottom sediments, although the radioactivity remained below hazardous levels. Much of the radiation was derived from the slight leakage of pressurized water reactors in nuclear submarines and from shore-based discharges of nuclear contaminants into the Thames River by Navy waste facilities.[92]

Groton was cited again in June, when a different study found that many of the Navy's best and brightest nuclear technicians were leaving the service to start families as a result of inadequate base housing. Captain McGuiness, a commanding officer at the base commented in a speech, "It isn't pay that forces him to leave, it's the inability to find housing close to the naval base's services, where his wife and children can be cared for, that forces him to get out."[93]

Elsewhere in New England, people could—and did—take greater pride in local operations. The *Boston Globe*, hailed MITRE, along Rt. 3 in Bedford as a guardian of the free world. The non-profit's original purpose of integrating SAGE with the Continental Air Defense System had given way to new goals. In 1968, MITRE's 2,000 workers labored to integrate the Back-Up Air Defense System, the Base Air Defense System, and the NATO Air Defense System, while coordinating a civilian air traffic control with the FAA.[94]

As the fantastic Apollo program got underway, New Englanders had another reason for pride. The bite-size, dehydrated foods that Apollo crews took to space were developed at Army Natick Labs. Food scientists at the Army facility had adapted the freeze dried, 2300 calorie nourishment to be ready instantly with 150 degree water available onboard the spacecraft.[95]

Even as Natick assisted astronauts in exploring outer space, the Labs also aided soldiers sent into the extensive networks of Viet Cong tunnels, which honeycombed South Vietnam. Special ear plugs, twin .22 caliber pistols, twin waterproof flashlights, and a special lightweight communication kit were put together for soldiers sent on the unpleasant task of clearing the tunnels.[96]

1968 witnessed few serious air accidents in New England, but several major air disasters in other parts of the world tied back to the region. At the end of September, a

KC-135 from Loring slammed into the runway at Wake Island in the middle of the Pacific, exploding and claiming 11 lives.[97] On October 2, two airmen from New England, plus other crewmembers aboard a KC-135 home based at Westover were killed when the aircraft crashed into a hill in Taiwan, en route from Thailand's U Tapao air base.[98]

The capture of the US spy ship *Pueblo* in 1968 turned into a media circus, one which circled around Ralph McClintock in May, 1969. McClintock was a Milton, Massachusetts native who had been onboard when the ship was captured.

Upon his release, McClintock told his story to the *Boston Globe*, telling of intense interrogations and frequent kicking and beating (often with the butts of machine guns) by the North Koreans. At one point, a major interrogating McClintock asked him to draw a sketch of Massachusetts and point out the state's military bases.

"This just stunned me. This was totally ridiculous. I almost broke out laughing. I didn't believe it. They looked very serious." McClintock sketched a map, showing the Boston Navy Yard, but omitted Hanscom, South Weymouth and Westover. "They knew about Westover and Hanscom. They wanted to know what kind of aircraft was at Westover and I told them it supplied food to Europe. They switched from that and accused me of working at Hanscom." The major drew out his pistol and threatened to shoot McClintock as spy for lying.

"I told him again I had never been to Hanscom, that I had no reason to go to Hanscom and that I knew not one lousy thing about Hanscom Air Force base. I told them the only thing that I knew about it was that it supplied Europe, too."[99]

In home waters, trouble continued with the Soviet trawlers. But in 1968, after several years of falling fish price and competition with the foreign fleets, crews at Boston's Fish Pier began to see a turn around. Starting in the early '60s and continuing until 1968, 65 percent of fish eaten in the US was imported. The culprits were actually the Canadians, this time, relying on government subsidies to undersell American crews, but the Soviets continued to receive the full blame. In fact, countries were not the only problem—the Catholic church had recently released its followers from a prohibition on red meat on Fridays, a requirement which had driven fish sales in New England for decades. With money from the federal government, the Massachusetts Seafood Council launched a campaign to raise interest in eating fresh fish.[100]

1969 saw a shake up on the home front. SAC was scheduled to leave Westover for good, along with 1150 personnel by June, 1970. SAC would still keep a foothold in Western Massachusetts, but most of its aircraft and all of its command and control duties would pass to Barksdale AFB in Louisiana.[101] At the end of the year, South Shore residents were alarmed to learn that Congress was considering moving South Weymouth operations to Quonset Point. Bridgewater Congressman Hastings Keith lodged a complaint, explaining that Quonset Point would not have enough space to house 3600 new reservists.[102]

On the home front, the late '60s brought a proposed expansion in the missile sites around New England, and 'Stop ABM' bumper stickers began to appear on cars throughout the region. The installation of BOMARC at Otis in the late '50s along with numerous NIKE missile sites, and Air Defense Command bases gave New England cities decent protection against Soviet bombers, but offered no protection at all from Soviet ICBMs. In the late '60s, the US and USSR both experimented with anti-ballistic missile systems—missiles to shoot down other missiles.

Initially, the US modified existing NIKE missiles, creating the NIKE-Zeus variant. Unfortunately, Zeus trials proved the system to be extremely ineffective. After experiments with different missile systems offered more promise, Secretary of Defense Robert McNamara proposed the Sentinel missile defense system, which he claimed would offer the US a nationwide blanket of protection against incoming ICBMs.

When Sentinel was proposed, McNamara insisted that the new ABM system would be used to defend against Chinese missiles. The system, he claimed would be overwhelmed by numerically superior Soviet missiles. Although McNamara's assessment was probably correct, there were also political reasons for defending against the "Chinese threat," which had hardly figured into US defense planning before. ABM systems were causing large amounts of international controversy with the Soviets. If an ABM system could be made to work effectively it could make a nuclear war "winnable."

On November 3, 1967, plans were released for the first 10 Sentinel missile sites. The Greater Boston area was seen as a high priority region for the project and the Air Defense Command went ahead examining different sites for an ABM missile silo. In August, 1968, a property at Sharpner's Pond in the suburb of Andover, Massachusetts was bought up. The site aimed to take advantage of the proximity of a National Guard facility where a missile acquisition and tracking radar station was also slated to be built. Congressional approval for the site came in September, and construction was well underway by January, 1969, when a Dracut-area contractor finished clearing the site, making way for an Idaho-based contractor to finish the work of building the silo.

The 1968 presidential election had turned ballistic missile defense into a hot topic and a major political issue, and turned Sharpner's Pond into a lightning rod for anti-ABM fervor. In spite of a blizzard, over 1,000 people arrived at Reading High School for an Army Corps of Engineers "town hall" meeting on ABM. The meeting was the beginning of the end for Sharpner's Pond. Soon afterwards, petitioners and letter writers managed to convince Senator Edward Kennedy to turn his attention to blocking ABM operations in the Boston-area.

Acting in Congress, Kennedy threatened to shut off funding for the Sentinel project if the new Secretary of Defense Melvin Laird failed to conduct a thorough review of the project. By March, the review had been carried out, with the Department of Defense concluding that ABM missiles would better serve the country protecting ICBM fields in the west, rather than safeguarding major cities. On March 14, 1969, work ceased at Sharpner's Pond and contracts with the different site developers were cancelled. The Sentinel ABM system was never constructed, and Sharpner's Pond was put to other uses.

The '60s ended in activism and anti-war unrest rather than all-out war with the Soviets, but the once peaceful cities and towns of the region had no idea what the next decade held in store.

# Atlas Missile Bases: Vermont Vanguard

A typical 556th Strategic Missile Squadron Atlas F-missile runs through its pre-launch procedure at a silo in Au Sable Forks, New York, near Plattsburgh in 1962. Unlike later missiles, such as the Titan, the Atlas had to be elevated out of its silo to launch.

Atlas No. 84 rises from its silo in Alburg, Vermont for a fueling test in 1962, in this image taken the Air Force.

The 556th Strategic Missile Squadron patch shows the unit's motto: "prepared for either"

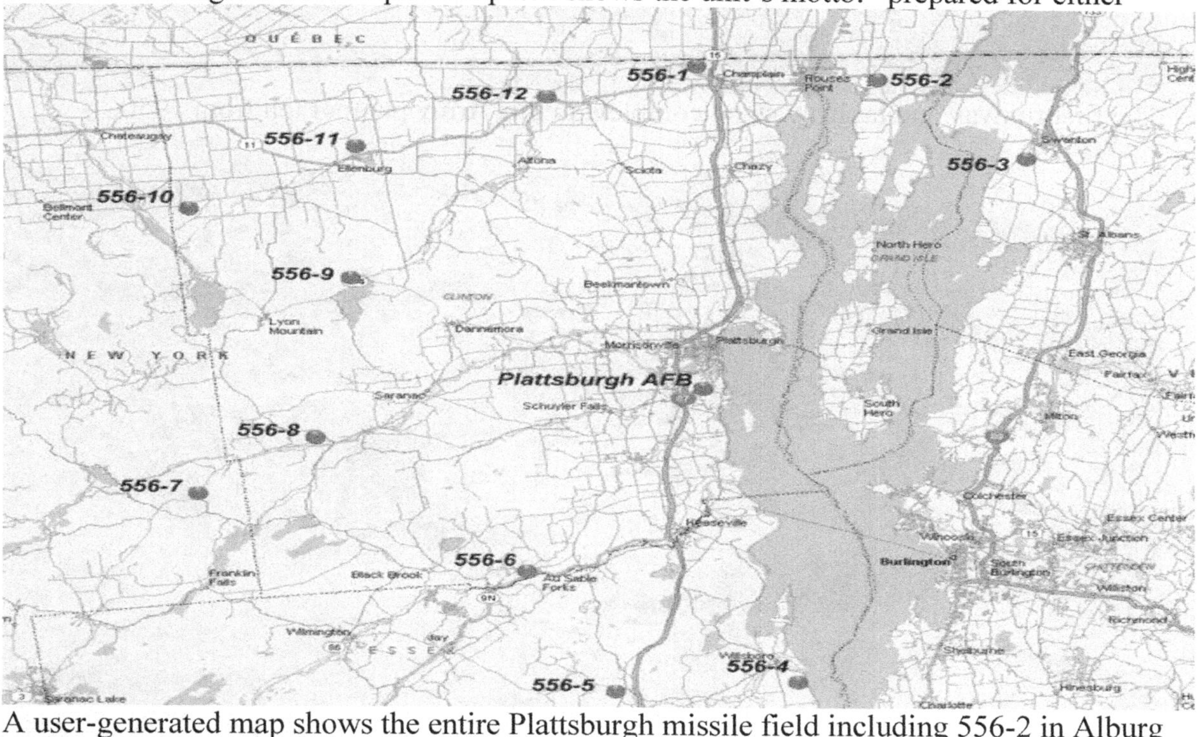

A user-generated map shows the entire Plattsburgh missile field including 556-2 in Alburg and 556-3 in Swanton, Vermont.

# Loring Air Force Base: World's Largest Air Base

The first runway and taxiway stand complete in this Army photograph from 1950.

Seen here in an Army Corps of Engineers photograph is the first double-cantilever hangar at Loring constructed in 1956. Note the B-52 parked on the tarmac in front of the hangar.

Construction workers pour concrete for a new runway in 1958, with B-47 bombers in the background. Loring was a source of employment for many civilians in Aroostook County.

A photographer records the work on an arch-ribbed concrete hangar in the late 1950s.

This aerial photo from 1970 shows nearly 30 B-52 and KC-135 aircraft on the apron—what may have been a slow day for the base. (Courtesy Library of Congress)

This oblique aerial photo shows the weapons storage area at neighboring Caribou Air Force Station in mid-1960s. The massive weapons igloos in the picture were designed to shield nuclear weapons from prying eyes, enemy attacks with conventional explosives, and the incredible cold of the Maine taiga. (Courtesy Library of Congress)

Aircrews load CAPTOR torpedoes onto B-52s on November 1, 1989 for a joint exercise.

A National Park Service image from 1999 shows a stretch of abandoned weapons igloos at the base. The same year, tens of thousands descended on Loring for a Phish concert.

# Quonset Point Naval Air Station: Coastal Outpost

Massive repair hangars take center stage ahead of an anti-submarine warfare aircraft carrier at Quonset's main wharf.

Workers piece apart the flight controls on a Douglas A-4 Skyhawk attack jet.

A Dougas A-1 Skyraider lingers on the runway. A propeller-driven anachronism in the Jet Age, Skyraider served with distinction in a Vietnam-era ground attack role.

An SH-31 helicopter being painted in the Paint Hangar.

Transports and maritime patrol planes undergo servicing in one of Quonset's many repair hangars.

A technician prepares aircraft survival rafts (above) while a courier races through Quonset's offices on a tricycle.

A-4 Skyhawk carrier-based attack jets undergo fuselage and engine repairs, paired with Douglas A-1 Skyraiders (above) and Grumman E-1 Tracer airborne early warning aircraft (below).

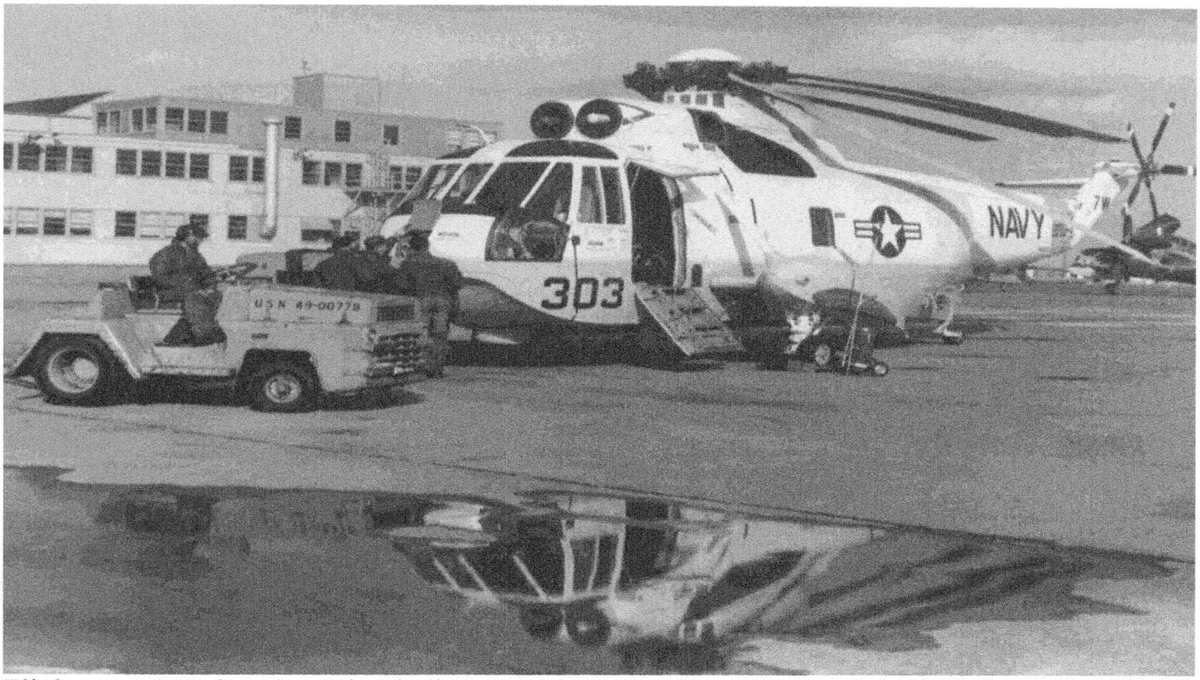


Flight crews ready a Sea King helicopter (above) and a Grumman Tracer (below) for patrols.

In a scene more reminiscent of Logan than Quonset, this ground crewman hauls cargo across the runway (above) while air traffic controllers scan the tarmac from the control tower (below).

A Grumman E-1 Tracker turns on the runway, unfolding its wings for takeoff.

Naval Air Station security heads out into Narragansett Bay (above). A team of engineers and mechanics gather in front of an A-4 Skyhawk, seemingly celebrating repairs that would extend the plane's life well into the 1970s.

A machinist inspects the heap of shavings by a Cintimatic drill press (above) while white-suited technicians staff Quonset Point's clean room (below).

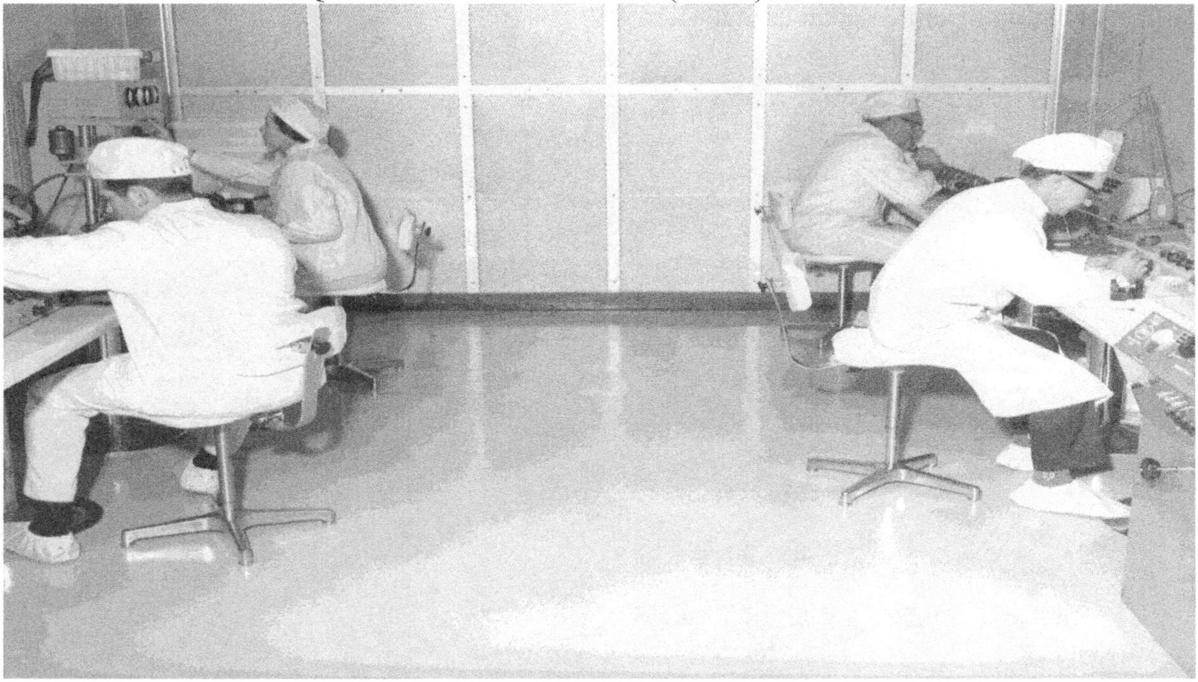

Quonset Point's wharf hosted anti-submarine warfare aircraft carriers, while the base served as the headquarters for US Antarctic operations. Below, a USAF Lockheed Constellation disembarks an unidentified VIP.

An unidentified sentry stops motorists at the entrance of Quonset Point.

Quonset's BOQ—Bachelor Officer Quarters—was the center of off duty life for unmarried officers.

Utility trucks and a coupe sit idle amidst a light snowfall at the station hospital.

Like the BOQ for officer's these barrack served as the living quarters for enlisted men.

Shown here in the late 1960s, the administration building was a hub of all base activities.

Perhaps more entertaining than the barracks, the Seair Club was central to Quonset Point social life.

Among the other perks of base life were the theater and Navy Exchange, pictured here in the late 1960s.

## Chapter III: The Uncertain '70s

The 1970s began with the ongoing war in Vietnam escalating rapidly while the antiwar movement grew increasingly vocal. 1970 brought a US invasion of Cambodia that infuriated antiwar figures. A protest against the invasion at Ohio's Kent State on May 4, led to the shooting of four students by National Guard troops. In a controversial move, opposed by most of the state legislature, Governor Sargent ordered flags lowered to half-staff after 15,000 protestors gathered at the Massachusetts State House.[103]

Throughout 1971 and '72, the Nixon administration grew more and more frustrated with the antiwar movement and struggled to muster support for the war. Although the war was in its final years, fighting continued heavier than ever with hundreds of thousands of Americans on the ground, and tens of thousands of bombs dropped each year.

The '70s were characterized by a continuation of the 1960s riot culture in urban areas. On July 10, 1970, riots broke out on the South Coast of Massachusetts in New Bedford. All three shifts of the New Bedford police and an additional 25 state troopers called in from Middleboro massed in the city to battle rioters in the West End. The riots began at 10 p.m. and quickly produced serious damage. Police and firefighters were unable to reach a burning building on Kempton Street due to the presence of rioters. Elsewhere in the city, a two-story building near police headquarters was consumed by flames. Several buildings were damaged and a Masonic hall burnt to the ground.

Throughout the night, police were pelted with rocks and dodged fire from small arms. One area consumed by flames was cleared out with tear gas so that it could be reached by firefighters. However, the fire department had to withdraw when the police left to tackle rioters elsewhere in the city.[104]

An editorial published at the time of the New Bedford riots summarized the struggle behind the violence. New Bedford was a stratified city in the '70s, with a stark divide between the poor, predominantly black residents of the Asbury Park neighborhood and their wealthier white neighbors in other parts of the city. In the late '60s 1200 units of low-income housing had been torn down to make way for highways, and employment in the city stood at 8.3 percent. As the author of the editorial put it, "In Asbury Park, the blacks finally had their fill of joblesness, intolerable housing and general indignities in the ghettos into which they had been crowded for generations as servants of summer people...[they are] no longer willing to be seasonally adjusted."[105]

"They're ready to die in the streets," commented one official about black youths in New Bedford, who faced a 32 percent unemployment rate and dim prospects on the road ahead. Mayor George Rogers managed to quell the violence with local police, rather than the automatic weapons and tear gas of the National Guard by imposing a curfew on the entire city. Twelve people were arrested for violating the curfew.

Unfortunately, the curfew came too late for 17-year old Lester Lima on Kempton Street. A group of men drove by a street corner, and gunned down the black teenager with a shotgun, fatally wounding him. New Bedford police managed to capture and arrest the three white killers responsible for the crime. Nonetheless, New Bedford's black population erected barricades in the streets as the curfew took effect, fearing reprisals from white residents.

A council of five African-American community leaders met with the police, explaining the need for the barricades. One of the group told reporters, "[It's] to keep white men out of the area. We want peace but we have to have barricades."[106]

Unfortunately, New Bedford was far from the last protest—or riot—in '70s New England.

On May 14, 1972, 1500 antiwar protestors marched through the streets of downtown Boston and across the bridge to Charlestown, proceeding to the Charlestown Navy Yard in anger at the US blockade of North Vietnam.

Fran McPoland, a leader of the Student Mobilization Committee told reporters, "We really want to demonstrate against the Navy, because this is the Navy's war now." At University of Vermont and MIT, students vacated classrooms in ROTC buildings throughout the day as a show of protest.

The protestors were joined by members of the High School Mobilization Committee, the Communist Party, the Boston Women's Abortion Action Committee, and the Socialist Workers' Party. They were jeered by Charlestown locals as they made their way to the Navy Yard. "Where's the grass?" queried one man at a cafe, flashing the protestors a peace symbol.

"Why don't youse all go home? Go back to Cambridge, why don't youse? Charlestown takes care of itself," said another man.[107]

At the end of the month, the Smith College graduation ceremony was clouded by antiwar protests. Thirty graduating seniors wore antiwar armbands as part of their graduation vestments. One student wore a sprig of flowers, a US flag, and a sign across her back that read, "Stop Murder." One hundred fifty students and faculty members gathered at the gates of Westover in the early evening for a service. A dozen of them sat down at entrance and waited for the Air Force to arrest them.[108]

1972 was the beginning of the end for the Vietnam War, although it was still a year of intense fighting in Southeast Asia. Elsewhere in the world, Portugal, a country that could easily fit within the confines of New England, with plenty of space left over, was embroiled in its version of Vietnam—or more accurately, three of them.

At a time when the kind of globalization that we know today was just beginning, with container ships plying the ocean, and jet planes passing quietly overhead, Portugal clung to a past that was out of step with the rest of the world. Portugal had been the first European country to set up colonies in Africa in 1400s and 1500s, and in the '70s, it was the last colonial power standing, with huge tracts of territory such as Angola, Mozambique, Guineau-Bissau, and the small island chains of Sao Tome and Principe and Cabo Verde (better known in America as Cape Verde).

The brutal Salazar military dictatorship conscripted hundreds of thousands of young Portuguese men to fight in a last ditch effort to save its colonies from communist-led independence movements. In southern Africa, the Portuguese efforts were largely successful in Angola and Mozambique, but in the small tropical colony of Guinea-Bissau, rebels crowded the Portuguese into only a few cities and towns.

Historically, Massachusetts has always maintained close ties with the seafaring peoples of the Azores and Cabo Verde, who together staffed the fishing fleets and populated the neighborhoods and factories of cities such as New Bedford, Fall River, and Brockton.

In 1972, a New Bedford sailor known as Pink was pressed into government service, working for the CIA. Using one of the many small freighters sailing out of New Bedford, Pink loaded up firearms and explosives by the ton. None of the men in Pink's team had any experience working with explosives, and the CIA officers supervising their work refused to offer any guidance.

Once the freighter was loaded, Pink and his men would sail across the Atlantic, making stops in the Azores, Cabo Verde and Angola to offload weapons for Salazar's troops. In tight-knit New Bedford fishing circles, word soon spread, and the locals were not happy. New Bedford's Azoreans and Cape Verdeans all despised Salazar.

Both groups hated that Pink's shipments were going to support a war that their relatives in the home islands were being drafted to fight. Both groups blamed Salazar for exploiting the economies of their beautiful islands.

Thievery is not uncommon in ports, and New Bedford was no different. Hundreds of grenades and other weapons disappeared from Pink's freighter. Although he did his best to stop the loss, going to the authorities was out of the question. Even if they were stealing, Pink didn't want to see any of his men go to prison. Things reached a head as New Bedford's Cape Verdeans caught on. A small group of Verdean radicals threatened to blow up and sink Pink's freighter in New Bedford harbor if he kept shipping weapons to the Portuguese.

Pink managed to calm the radicals, getting them 'onboard' for his missions. If they helped to ship the weapons to Africa, he would clear space for them to bring furniture and supplies to their relatives back in Cabo Verde.[109]

At sea, weapons trafficking was not the only thing troubling New England mariners. On November 23, a horrifying incident galvanized public feeling against the Soviet trawlers once and for all. During a Coast Guard inspection, Simonas "Simas" Kudirka, a Lithuanian sailor, leapt off the deck of the mother ship *Sovetskaya Litva* and came aboard the cutter *Vigilant*, in a bid for political asylum. The *Vigilant*, sailing out of New Bedford (and in still in service in the 2010s) was ill-equipped to deal with the incident, even though they were moored in American waters. Kudirka stayed aboard for 10 hours as the crew attempted to raise the State Department on the radio.

Unfortunately, a radio anomaly prevented the *Vigilant* from contacting receivers in Marshfield, Massachusetts. Meanwhile, the Soviet ship's officers accused Kudirka of stealing 3,000 rubles and demanded that he be returned to face justice. When the *Vigilant* finally raised Admiral William Ellis, onshore, they were ordered to let a KGB detachment onboard.

At 10:14 in the evening, the *Vigilant's* highest ranking officer, Commander Ralph Eustis called his ranking officer, Captain Brown at his house. Eustis proposed a last ditch effort to protect Kudirka, suggesting that he bring the 40-year old Lithuanian back to New Bedford to allow the State Department to sort out the situation. Brown interjected, "Commander Eustis, you have your orders, you have no discretion. Use whatever force is necessary [to hand over to Kudirka]."

In a later interview, Brown admitted what happened next. "When I hung up the phone, why, well, I guess I totally broke down in this house with my wife...I told my wife... 'I spent all my life saving people, and I finally had to give an order, and I think it's for a man to be killed." Bowing to his orders, Eustis allowed three KGB officers assisted by three other sailors to force open the door of the head (the shipboard bathroom), where Kudirka was hiding. The KGB officers began kicking and punching Kudirka immediately, although he quickly slipped away and sprinted down the narrow passageway, shouting, "Help me! Oh, God help me!"

Robert Brieze, a Latvian refugee and president of the New Bedford Seafood Producers Association attempted to intervene but was ordered aside. The KGB officers caught up to Kudirka on the *Vigilant* and slammed a door on his hands. He escaped again, hanging onto the lip of the hull by his fingertips for a few moments before slipping back onto the deck. He was blocked by two American coastguardsmen who grabbed hold of him and held onto him until the KGB turned up. One of the Soviets attempted to choke him with a telephone before hauling him up a ladder to the *Vigilant's* helicopter pad, smacking Kudirka's head into each of the rungs on the way up. The Soviet's continued to beat and attempt to strangle Kudirka until the American crew intervened. He was bound with ropes and a blanket and dropped three feet onto the lifeboat's engine hatch, landing on his head.

The American crew piloted the lifeboat a mile to the mother ship, while the KGB officers and sailors subdued Kudirka by sitting on his head and punching him repeatedly, as the Lithuanian slipped in and out of consciousness. "If he is not dead now, he will be dead in a few minutes," one of the Soviet officers told the Americans ominously. At the mother ship, Kudirka was brought aboard with a rope and tire hoist. The captain of the vessel lowered down a gift for Commander Eustis—a bottle of Johnnie Walker Red Label.

The story broke the following day in the *New Bedford Standard-Times* and was quickly picked up by the *New York Times*. A *Washington Post* editorial summed up the situation, "No more sickening and humiliating an episode in international relations has taken place than the American government's knowing return of a would-be Soviet defector to Soviet authorities on an American ship in American waters."

In subsequent investigations, Commander Eustis was transferred to shore leave and Captain Brown was forced into retirement. Kudirka faced a bleak future back in the Soviet Union. Upon his arrival he was sentenced to 10 years in prison for treason, and remanded to Potma prison in Moldova. After going on hunger strike, Kudirka was reassigned to the deadly Special Correctional Work Institution in Perm at the end of Siberia, to face terrifying winter temperatures and grueling 'correctional' work. Several years after the Kudirka incident, a Lithuanian immigrant in New York discovered that the erstwhile defector might be able to claim US citizenship through an American mother. With Kudirka's case gaining traction, Secretary of State Henry Kissinger lobbied for his release through the Soviet embassy. At the end of 1974, Kudirka was released from prison and relocated to the US along with his family, to become a janitor in the Bronx. Kudirka ultimately returned to an independent, post-Soviet Lithuania in 2002.[110]

While the Kudirka incident made waves onshore, fishing crews dealt with the fallout of 1960s overfishing. With most of the richest fishing grounds in international waters foreign trawlers proved incredibly efficient at removing *all* of the fish from the sea. The Gulf of Maine, one of the most productive fisheries in New England was the main source of income for thousands of Maine residents, but by the mid-'70s it was nearly fished out.

The initial foreign arrival had increased competition for fish sixfold according to Maine resident and journalist Colin Woodard. US fishermen caught 556 million pounds off of New England in 1968, compared with 1.2 billion tons caught by the foreign fleets. The ravenous factory ships were incredibly wasteful beasts. Each trawler's net would scour the sea floor clear of all life and then turn its catch over to a factory ship or process it on board. Ships specialized in certain types of fish and discarded other species caught in the nets. Tons of dead fish were dumped overboard, left to rot and pollute the bottom.

The Soviets and other fishing fleets would focus on one species at a time, harvesting it until few fish remained and then moving onto another. The productive Narragansett Bay lobster industry was scandalized when one lobsterman recovered a catch with a Cyrillic tag painted on its shell, evidence of Soviet interest in the lobster population. Although the Soviets never pursued lobsters for their fleet, the impact of foreign ships put mounting pressure on US fishermen to simply survive.[111]

In response to foreign overfishing, of the East and West Coast of the US, Congress passed the Magnuson-Stevens Fishery Conservation and Management Act, better known as the Magnuson-Stevens Act in 1976, with sponsorship from senators Magnuson and Stevens, from Washington and Alaska respectively. The Act set up eight special councils to regulate fishing and creating a US 'Exclusive Economic Zone,' that pushed the territorial limit 200 miles out to sea. Unfortunately, the Magnuson-Stevens Act remained nearly toothless until a 1996 revision. Although the Act did curb some foreign fishing, Soviet trawlers continued to arrive in large numbers throughout the 1980s, driving the destruction of New England's fisheries while American crews caught the fish that remained.

The Paris peace accords and the American withdrawal from Vietnam in March, 1973 coincided with the end of the draft. The war had claimed the lives of 58,000 American service members and left both North and South Vietnam in shambles. In addition to deploying herbicides such as Agent Orange, the US had dropped almost three times as many bombs on Vietnam as they had on Germany and Japan combined, during World War II.

The end of the war did not end the protests. In fact, acts of civil disobedience grew among military personnel. In the aftermath of the Paris peace accords, forward deployed bombers from Westover continued to bomb the Cambodian countryside throughout much of 1973. Captain Donald Dawson, a bomber pilot from Westover deployed to U Tapao air base in Thailand, refused to fly on a combat mission in July, 1973. Dawson faced the possibility of a court martial and dishonorable discharge from the military, or even a long stint in military prison for disobeying his orders.[112]

Dawson was not the only opponent of the Cambodian air war within the Air Force. Only a few weeks later, Captain Michael Strain, an Oklahoman bomber pilot stationed at Westover launched a class action lawsuit in a federal court to end the bombing.

"This suit was not a personal thing with me. It's more of a citizen's action," Strain told reporters. He added that he would "definitely" fly combat missions against Cambodia if ordered to.

On July 25, a federal judge ruled that the bombing in Cambodia was unconstitutional. "There is no existing congressional authority to order military forces into Cambodia or to release bombs over Cambodia." Helping Strain, and two other pilots stationed in Guam to arrange the lawsuit, was Representative Elizabeth Holtzman, a Democrat from New York.

In addition to being one of only a few women in Congress at the time, Holtzman asked to be listed as 'Ms. Holtzman' in news stories, using what reporters called the "feminist designation" (Ms. was only beginning to become a common term of address in the '70s).

Holtzman was thrilled with the judge's decision. "It reasserts the constitutional requirement that no American lives can be sacrificed, that no American lives can be risked, and that no American taxpayer's money can go to a military effort that has not been approved or authorized by Congress," she announced happily.[113]

Americans were already turning their attention to a different part of the world, as the Middle East supplanted Southeast Asia in the news. In October, 1973, Egypt, Syria and other Arab nations launched a surprise attack on Israel during the major Jewish holiday of Yom Kippur in hopes of retaking the Palestinian territory occupied by the Israelis since 1948 (other territories such as the West Bank and Gaza Strip had been added during a similar Israeli operation in 1967). Israel won the war, but only narrowly. For a brief instant, it appeared that the young nation might get swept away.

America intervened, sending weapons to resupply the Israelis. In revenge, the Arab nations belonging to OPEC (the Organization of Petroleum Exporting Countries) imposed an oil embargo on the US, increasing the price of oil from three dollars a barrel to 12 dollars. The shock set off the 1973 energy crisis, which soon resulted in chaos as gas stations ran out of fuel for gas guzzling American vehicles. Rationing and 55 mile per hour highway speed limits were soon to follow.

To add insult to injury, Americans learned in 1975 of an oblique US defeat in Vietnam, when South Vietnamese forces in Saigon were overrun by invading North Vietnamese troops. The communists had triumphed in Vietnam and the same year swept into power in Cambodia and Laos. Thousands of South Vietnamese fled, fearing execution or long imprisonment in 'reeducation camps.' Although many of these Vietnamese settled in California and Louisiana, many others arrived in Massachusetts to start new lives.

Nguyen Van Thieu, the last president of South Vietnam was part of the exodus. Telling his countrymen half-heartedly, "I resign, but I do not desert," Thieu left the country for Taiwan aboard a CIA flight and then traveled to Britain. In the early 1990s, he joined the thousands of Vietnamese refugees who had settled in New England, taking up a reclusive lifestyle in Foxborough, Massachusetts. He ultimately died in 2001 at Boston's Beth Israel Deaconess Medical Center. By then, Lowell, Massachusetts had become the center for one of the largest Southeast Asian immigrant communities, as thousands of Cambodians settled in the city. The large Cambodian population had arrived in the aftermath of Pol Pot's horrific genocide in the late 1970s, which ironically, the Vietnamese communists had ended by invading Cambodia in 1978 (Vietnam's involvement in Cambodia turned into a replay of America's involvement in the Vietnam War, with Vietnamese troops struggling to fight Pol Pot's Khmer Rouge rebels).

Perhaps one of the strangest stories from Cold War New England is the tale of the cockroaches that ate Natick. US Army Natick Labs was one of the smallest Army facilities in New England, but nonetheless, one of the most important. With a focus on research and development, some rather whacky ideas emerged from Natick in the '60s and '70s. For instance, scientists at Natick labs designed combat boots with rubber soles that left tracks identical to a Vietnamese peasant, in the hopes of hiding the movement of US patrols from curious Viet Cong fighters.

Other strange ideas to emerge from Natick included "sunscreen" to protect soldiers from nuclear blasts. The idea was tested on a group of unfortunate pigs who were cooked alive in a solar furnace in the name of science. In another moment of scientific curiosity, the lab imported large numbers of Madagascar hissing cockroaches in the early '70s to see how long they could last when exposed to large doses of radiation.

As it turned out, the roaches were more durable than the researchers had guessed. When the experiment ended in 1974, the roaches—many of them still alive—were bagged up and doused in carbon tetrachloride before a trip to the Natick dump. The carbon tetrachloride killed the adults, but left the eggs intact, and ultimately burned through the bags, unleashing a great many hungry baby roaches on unsuspecting Natick homeowners. The town health inspector did his best to calm the locals, but gave up when residents threatened to stuff his mailbox with a delivery of roaches. The Army, the town and the residents spent nine months dealing with the problem, with the Army giving out buckets of DDT to residents.

The fall of 1974 witnessed a return to violence in Boston, with a major riot in Roxbury. The incident in Roxbury was essentially a race riot between black and white residents. Twenty four whites and 14 blacks were injured during the incident.

The riot started outside English High School in Roxbury. From there, it quickly spread throughout the Mission Hill and up and down Huntington and Longwood Avenue.

The violence was just the beginning of the extreme racial tension witnessed in Boston during the 1970s busing crisis. In a well-intentioned, but ill-starred plan conceived in part by liberal suburban residents, living on the outskirts of Boston, and by federal mandates, the city's predominantly black and predominantly white neighborhoods would be integrated by busing students between schools. Unfortunately, integration was forced through in a heavy handed way, mandated by official court orders. The Roxbury riot marked the 19th day of the integration plan.

The start of widespread busing led to violence on both sides.

Black youths roamed the streets of Roxbury smashing car windshields with rocks, and setting at least two vehicles on fire. The city's police commissioner imposed a temporary prohibition on alcohol sales in South Boston and Roxbury after a black motorist was dragged out of his car and savagely beaten by a group of whites in South Boston. Police managed to intervene and save the man's life.

Bostonians were confused and scared by the level of violence in the city. Throughout the Boston school system, attendance for all grades stood at just 69 percent. Many universities such as Boston State College (later, a division of UMASS-Boston) and Wentworth Institute of Technology cancelled evening classes. At Hyde Park High School, a boycott by black students followed the beating of the black motorist in South Boston.

"All we can do is our best," explained one police officer.

On October 10, 1974, only a few days after the Roxbury riot, Mayor Kevin White deployed 320 state troopers, and an additional 125 Metropolitan District Commission (MDC) police to protect black students at South Boston High School and the Gavin middle school.

According to reporters, some of the MDC police "were in battle dress." They wore paratrooper jumpsuits and carried gas masks in their utility belts. One fifth of the entire MDC force was deployed to South Boston. The State Police on the other hand wore crisply starched dress uniforms.

The police made efforts to "diplomatically" disperse small crowds throughout South Boston. The state and MDC police had been given the job of filling in for the controversial Boston Police Tactical Patrol Force, which had featured significantly in the early rioting.

However, some observers remained worried. "It's the same relaxed, professional air the Tactical Patrol Force had Sept. 12, opening day at South Boston High School. They, too, were diplomatic in moving groups of people, and their orders were obeyed with some good grace. But that was almost a month ago."[114]

While MDC and state police attempted to keep the peace in South Boston, Massachusetts Governor Francis Sargent spoke to President Ford about getting federal troops to prevent race riots in Boston. The White House released a diplomatically worded statement, "[no federal assistance] would be in order until the governor is in a position to say he has utilized the full resources of the state and that despite these efforts he can no longer control the situation."

President Ford was quizzed about his decision on the way to give a speech in Kansas City. Ford told reporters that he deplored the violence and added, "[I believe] the main responsibility for maintaining order continues to lie with the state and local officials."

At the time, Sargent's administration was not actually in a position to request federal troops. Department of Defense protocols called for using federal troops only in rare situations when the National Guard and local police became overwhelmed. During the contemporary Detroit riots, 10,000 Michigan National Guard troops had been joined by 4700 federal soldiers, when rioting got out of hand.

Sargent had activated the ground units of the National Guard, but was holding them in reserve in state armories in case extreme violence struck again. The Department of Defense was not sure which troops it would send, if matters worsened. The 82nd Airborne at Fort Bragg, North Carolina were seen as likely candidates in the fray. Federal troops had been deployed throughout Chicago, Baltimore and Washington during the Martin Luther King assassination riots and anti-war protests in Washington.[115]

Later in the month, South Boston was rocked by another disturbance. This time, white supremacists descended on Boston in a show of force. Police arrested Matt Koehl, the 39 year old leader of the National Socialist White People's Party in City Hall plaza in the midst of Koehl's speech calling for an all-white nation. Boston's Deputy Superintendent of Police, Joseph Sala explained, "There was some pushing and shoving going on between Party supporters in the crowd and some hecklers. The crowd was also blocking the sidewalk and stopping traffic, so I ordered Koehl to be arrested as the person causing the disturbance."

After Koehl's arrest, a dozen neo-Nazis left City Hall in a yellow school bus labeled with a large 'White Power' sign. Three members of the group were arrested in South Boston, handing out flyers emblazoned with swastikas, on charges of inciting riot.[116]

Boston's busing crisis remained intense throughout 1975 and 1976, and lingered as a citywide political issue until the program ended in 1988. However, trouble was far from over in August, 1975, when 4,000 white residents and an additional 2,000 black residents gathered at Carson Beach in South Boston for a "Southie Pride Day." Six local priests had helped to arrange the event, and hailed it as part of an effort to "reaffirm the long-standing right of all people to use the beaches in South Boston."

After a slow start to the event in the morning, a group of whites wearing yellow armbands gathered along the beachfront, with MDC police interspersed. A small group of young black men attempted to make their way to the beach, but were turned back by the police, who feared that they would be injured by the angry, white mobs.

As small black groups coalesced and made their way down the sidewalk, they were taunted by whites riding by on bicycles, and even spat on.

Growing numbers of black residents gathered at Carson Beach, separated from white mobs by a thin line of MDC police officers, and 10 yards of sand. Off shore, Coast Guard boats zipped back and forth while helicopters hovered overhead.

Matters quickly got out of hand, as both mobs exchanged insults. Insults were quickly replaced with hurled bottles, shoes, and rocks. MDC police waded into the shallows with their horses in an attempt to prevent hand to hand fighting between the waders.

Ice cream trucks did a brisk business throughout the afternoon, in spite of large rocks striking car windshields all around them. While some people lined up to buy ice cream cones, others cowered in fear in the shelter of the trucks.

MDC police and the Boston's Tactical Patrol Force attempted to separate the groups, opening up 130 yards separating the two mobs. Some people regrouped on the road and continued to trade rocks, while police motorcycles raced by in an attempt to end the fighting. Skirmishes continued throughout the afternoon. A fist fight broke out, involving 20 people, while elsewhere on the beach a white girl was dragged into the water and her head held under water. The black mob that had caught up with her, pulled the girl's hair and shouted, "white bitch."[117]

In response to the ongoing racial tensions in the city, Michael Dukakis, the new Governor of Massachusetts, ordered National Guard troops into the city as the 1975-'76 school year got underway. The arrival of 600 troops in South Boston prompted a fresh wave of demonstrations. One police officer was stabbed in the chest, while a soldier was injured in the leg by a rock. The troops, drawn from four units of riot trained military police were sent into South Boston after 258 out of 775 police officers on emergency duty failed to report.

The guardsmen arrived in trucks from Camp Curtis Guild in Wakefield, at their new quarters in the Fargo naval building. Anti-busing demonstrators pounded on the sides of their vehicles as they passed. Other National Guard units from Waltham, West Newton, Dorchester and Braintree were activated and sent to Wakefield for emergency training, while Mayor White cajoled the Boston Police into mobilizing its entire force of 2000 officers. "The public safety forces must and will be adequate...The parents and children of the city must know and be confident of that fact," White told reporters.[118]

As the National Guard arrived, young people and adults throughout South Boston stoned cars, and shattered the windshield of at least one military van. Demonstrators stood outside the Fargo building throughout the day, shouting, "You stink," at the soldiers, while waving signs that read, "Cops or Troops, We'll Still Fight," and "Southie Says Forced Busing is Kidnapping."[119]

The arrival of skilled National Guard soldiers should have helped to quell the trouble in South Boston, but if anything, the trouble increased. In February, 1976, demonstrators hurled tear gas at the police and tossed trash cans off of rooftops. One officer, Robert MacAleese was temporarily blinded in both eyes. Reporters caught up with MacAleese on his porch, where he sat sightless below a ROAR flag (Restore Our Alienated Rights). Although MacAleese opposed busing, he told reporters, "I just wish they wouldn't throw rocks and bottles at police. We have a job to do and we're going to do it."[120]

Intense violence continued through the autumn of 1976. At the end of October, a pedestrian spotted a stick of dynamite with its fuse "sputtering" at Charlestown High School and summoned police. The same day, across town, an attempted firebombing damaged South Boston's high school. Six Molotov cocktails—improvised incendiary bombs made from glass bottles filled with gasoline—destroyed a home economics classroom. Outside of the school, police rounded up 23 protestors who were leading a protest march and throwing rocks through the school's windows.[121]

On April 19, 1976, word of a new Soviet documentary reached Boston newspapers. Soviet filmmaker Valentin Zorin issued *Boston Contrasts*, a 30 minute documentary that showed cheerful images of Boston's new skyscrapers and old Kennedy rallies that contrasted with the ugly scenes of race riots in the streets below. The documentary was screened twice in a month on Soviet television, describing Boston as an "arena of racial injustices." Zorin editorialized, "The young citizens of Boston learn not just multiplication tables and spelling rules..." Boston was described as a typical American city in which, "working people who create its glory and riches with their hands and a bourgeoisie which worships the dollar and the fist."[122]

The National Guard presence in South Boston lasted for years, as violence remained a perpetual threat. As Boston's most predominantly Irish neighborhood, there was a certain amount of irony to the situation. The 1970s were perhaps the most violent time in the Irish "Troubles," a period of widespread terrorism and insurrection in Northern Ireland. Many of South Boston's residents hated the British government almost as much as they hated the outside forces orchestrating busing. In fact, throughout the Troubles, much of the Irish Republican Army's funding came from New England Irish-Americans, many of them living in South Boston.

While South Boston was torn apart by riots and a de-facto military occupation, the relatives of Boston's Irish in Northern Ireland faced a more serious situation as British paratroopers searched house to house in neighborhoods obliterated by IRA car bombs. Many Irish-Americans in the Boston area supported IRA terrorists out of a misplaced sense of nationalism, believing that they were helping their people to fight for independence. In fact, the IRA's other main source of funding and training was the Libyan government, and the Stasi, East Germany's secret police. The IRA's goal after creating a united Ireland was replacing the island's democratic government with a new communist power structure.

The 1970s witnessed the high water mark of urban violence, but were also a time of widespread UFO activity. Whether the lights in the sky were actually visitors from another world, or not, what is clear is that many different people were seeing unexplained things in the '70s. In October and November, 1975, one famous incident unfolded at Loring.

Just before eight o'clock on the evening of October 27, Danny Lewis, a military police sergeant on duty at the weapons storage area heard a whirring noise overhead, much like a helicopter. The aircraft's lights came into view as it hovered 300 meters away from the weapons dump. Lewis raised the alert. Guards were deployed and the base's radar picked up the object on their scopes. The object continued to whir, emitting red orange light and a white strobe, as it circled the base for an hour and a half. The 42nd Bomb Wing commander requested emergency backup from NORAD F-106 fighters which scrambled from Hancock, New York and North Bay, Ontario. Loring wired the Pentagon and CIA headquarters in Langley—neither group could make head nor tail out of the mysterious lights in the sky.

The lights returned over the next three weeks. But they were not confined to Loring. Helicopter and fighter pilots tracked the mysterious objects together with radar operators on the ground at airbases and missile bases in Michigan, North Dakota, Montana and Ontario.

The same mysterious craft returned two more times to northern Maine. KC-135 tankers from the base raced across the skies of northern Maine, pursuing the UFO. Whether it was actually visitors from another world, or testing of a first-generation stealth fighter, we may never know.[123]

While Boston struggled with the specter of busing, bases and companies throughout New England faced ongoing spending cuts.

After the Nixon and Carter defense cuts, some good news arrived in November, 1977 from the Department of Defense. A plan was underway to revitalize military activity in the region. The New England congressional delegation was informed of three proposals, including one leading option that called for reopening Newport as a major homeport for the Navy. Other plans suggested reopening the Boston Navy Yard as a ship repair facility and relocating 33,000 troops from South Korea to Ft. Devens.

Deputy Defense Secretary Charles Duncan told the Congressmen, "We are re-evaluating putting surface ships in the First Naval District and that study should be completed by the end of the year." A re-opened Newport would allow the Navy to alleviate overcrowding at its other ports, by reintroducing a squadron of destroyers. In 1977, the Navy had 400 ships, with construction underway at Norfolk, Virginia and other shipyards to increase number to 560, leading to overcrowding problems there.

Congressmen from Maine were more concerned about Loring's future than Newport. "If Loring is closed down, there will be economic havoc in Aroostook County...It is the biggest industry there," Representative David Emery explained to the military.[124]

In 1977, Loring was still a very active base, although it was no longer quite the hive of action that it had been during the 1960s. Reporters visited the base to get an in depth look at the life that service members and their families led in Aroostook County. Readers back in Greater Boston were introduced to a cast of typical Loring characters, such as 32-year Louis Landry, who had quit a career path in agri-business in his native Louisiana to become a B-52 pilot in 1967.

By 1977, Loring was a well-established fixture on the Aroostook landscape, with an on-base population of 9,000 people, living on a 14,000 acre reservation. Although

ICBMs had changed the game since the '60s, Loring was still a strategic location that would make full use of its position, 300 miles closer to Europe than other continental bases, if a shooting war ever broke out.

In addition to the ever present threat of accidents and nuclear annihilation, residents at the base faced other challenges. Neighboring Limestone was still only a tiny community, with few diversions from everyday military occupations. For much of the year, Aroostook County is blanketed in thick snow, typically several feet deep, with temperatures that often plunge to 40 below zero. Adding to these hardships, the '70s were a time of particularly cold winters.

Loring hosted as many as 450 aircraft at a time in the late '70s, including 15 B-52s from the 42nd Bomb Wing and 31 KC-135s. Due to the recent 1973 energy crisis, bombers had been restricted to only three flights each month. The rest of the time, crews rehearsed for war with primitive flight simulators on the ground, as well as going through classes and briefings to help them get to enemy targets.

However, when fuel budgets allowed, the five man crews of the B-52s would spend two hours running through pre-flight checks before boarding their bombers for high and low altitude training over the Adirondacks, near Plattsburgh. Training missions typically involved mock attacks by Air National Guard fighter planes and in-flight refueling with the KC-135s. Every three weeks, a small group of bomber and tanker crews would be picked for alert duty. They would stay in Loring's base alert facility, within view of their aircraft, and within earshot of a klaxon.

Alert duty was a tense activity. Administrators had tried to ease the strain by allowing families to visit the aircrews in the alert facility. Nonetheless, one tanker pilot described alerts as, "the time when one of my kids gets sick or my car won't start and there's nothing I can do about it...You have to depend on the neighbors."

Many members of Loring's aircrews feared what would happen to them on the other side of civilian life if defense cuts continued. Many of them had sacrificed their prime working years, pursuing military occupations that did not always relate closely to civilian life. "What does a boom operator on a refueling tanker do? Run a gas station?" wondered one air crewman.

Many crews felt that they were being forgotten about by the American people. Bombers and tankers had never had much glamour to begin with, and during peacetime, there were no heroes in SAC. For aircrews made up largely of officers, the struggle to find promotions was intense. If an officer did not become a major after 10 years in the service, his future prospects in the Air Force would be limited.

Cynicism wasn't uncommon in SAC. Air crews nicknamed their B-52s BUFF, short for Big Ugly Fat Fellow. B-52s were sturdy airframes, but ungainly aircraft, and strange Air Force policies did not help very much. Every practice mission was required to be nine hours long, equivalent to a roundtrip flight to California. After evading "enemy" fighters and conducting in-flight refueling, crews often returned early to base, but were then forced to burn up fuel—and time—making touch and go landings to fulfill their full nine hours.

However, air crews never seemed to joke about their aircraft weapons. It was hard to make light of nuclear war, especially when it seemed that each of the air crews might be flying four nuclear bombs apiece to targets in the Soviet Union. On the frequent nuclear runs, the bomber crews simulated dropping aluminum chaff and Quail decoy missiles to confuse enemy radar, while making their final approach to drop their "strategic weapons." Each bomber could alternately haul 28, 750-pound bombs with conventional explosives—the type widely used in missions over North and South Vietnam.[125]

Although the Vietnam War was over, growing proxy wars between the US and Soviet backed rebel groups in Africa, threatened to entangle the US once again. Many analysts in the CIA and military believed that the US would soon get involved with the ongoing Angolan Civil War in southern Africa. If the US had become involved, bombers from Loring and Pease might have taken to the skies again to flatten villages held by the communist MPLA government.

But trouble emerged from another part of the globe, impacting Groton in 1979, with the overthrow of the Shah of Iran. At the time that the revolutionary government announced Ayatollah Khomeini as the country's new leader, around 250 Iranian submariners were stationed in Connecticut for training. The Navy denied reports that Marines in riot gear had cordoned off the Iranian barracks, but confirmed other reports. Sailors on the base were briefed to be on the lookout for violence and demonstrations from the Iranians, and 20 guards practiced closing off a classroom on the base in case classes got out of hand.[126]

Although the tension ultimately subsided when the Iranians returned home, the late '70s remained an uncertain time at Groton. As early as 1976, Richard Trott, a worker for an employment agency in the Groton area commented on the large number of highly-trained and experienced submariners leaving the service, noting 34 inquiries from submariners at his agency in the space of three weeks. Thirteen of the men had been officers, while 21 were technical workers. Trott found that most of the applicants were fathers in their late 20s with children between the ages of three and nine. Most wanted to spend more time with their families. But there may have been more to it than that. With detente in full-effect and Vietnam wrapped up, the same apathy and purposelessness that had already started to sweep through other services was starting to catch up with the nuclear Navy.[127]

Continued trouble on and around the base did little to assuage local doubt at the Navy's presence. Nine sailors from the submarine *Tullibee*, were caught smoking marijuana. The Navy's policy was to eject anybody caught smoking the marijuana, unless they admitted to smoking. In this case, the sailors admitted to their misdemeanor and were rewarded with "evaluation and rehabilitation" for their honesty. On future deployments, the men would only be posted to surface ships or shore stations.[128]

In June, 1979, *Woodrow Wilson*, a *Lafayette*-class nuclear ballistic missile submarine ran aground three and a half miles from New London. Fortunately, the submarine had been traveling at less than one knot, and didn't sustain any damage to its hull or reactor. In fact, it was able to reverse off of the sandbar under its own power.[129]

An earlier incident in 1975 had proved more worrisome, when a bomb threat was phoned in to the base. Naval police were ordered to conduct a hurried search of the diesel submarine *Clamagore*, where it was said that the bomb would go off in three minutes. The threat proved to be nothing more than a hoax.[130]

The crowning mess at Groton didn't come until 1980, when 400 personnel were evacuated when a mid-July fire tore through a print shop, thrift store, and storage area on campus. The fire was damaging but claimed no fatalities. Decay appeared to be largely responsible. The 50 year old building lacked firewalls, and as part of an outdated security measure had bars on the windows which limited firefighting. "There will be an extensive investigation, believe [me]," Raymond Grenier, the base's deputy fire chief told reporters.[131]

As a result of detente and the first round of intensive BRAC, New England saw fewer problems in the skies (save for UFOs of course) than it had during the '50s and '60s—that is to say, fewer spectacular crashes. Nonetheless, with large numbers of military aircraft still in the skies, the threat of accidents remained substantial. A swing-wing FB-111 bomber, en route to Loring spiraled out of the sky during a training mission just before Christmas, 1975, over the small community of Garfield Plantation, Maine. Fortunately, both the pilot and the navigator ejected safely and were airlifted out by a helicopter.

Local restaurant owner Clarence Michaud described the crash to reporters. "I seen it as I was walking up to the front door...It was off to the west, over a ridge. I saw a trail of white smoke and then an explosion from the impact, and then a great cloud of black smoke rose into the air."[132]

In April, 1978, an Air National Guard jet plowed into the rocky soil of Noman's Land island, the target range off of Aquinnah in Martha's Vineyard. The pilot, who had flown the F-100 Super Sabre out of Bradley Airport with the Connecticut-based 103rd Tactical Fighter Group, was pitched 200 yards from the crash site and killed. The Coast Guard deployed a boat, helicopter and plane to search the crash site, with South Weymouth tapped to retrieve the wreckage. "We usually fly four to 16 planes a week, depending on the mission we're trying to achieve," the head of the Connecticut unit explained, adding that the unit hadn't suffered a fatal crash in 11 years.[133]

The '70s were a bizarre and trying time for New England and America as a whole. For the first time in the Cold War, the enemy no longer seemed to be on the outside. Fissures between different ethnicities and political viewpoints broke out into violence and discord. As the US finished up its troubled operation in Vietnam, many elements of the country came to regard the US as the enemy—an overbearing government that had willfully sacrificed nearly 60,000 of its own men and women only to kill hundreds of thousands more Vietnamese. The late '70s were an even more jaded time, as New England descended into apathy and drugs; bases and industries shut down and disappeared for good. For some, the future was starting to look like a bleak place, but the Cold War wasn't over yet. The nation was calling and it was time for New England to answer once more.

The Cold War was ending, but the question of *how* it would end still remained...democracy and alliance or the peace of ultimate devastation.

# Pease Air Force Base: Peacemakers

The entrance sign for Pease in 1987, when the base remained an active SAC facility.

Seen here on a runway in California at Naval Missile Center, Point Mugu, in May, 1971, these two EB-47s were the last B-47 variants in service. 52-410 (on the right) flew to Pease in 1977 marking the last operational use of B-47s.

# Boston: Inside the Military Hub

This aerial photograph from the late 1950s shows the Atlantic Reserve Fleet at its moorings in South Boston. The large rectangles are 20 World War II-era aircraft carriers—more than the number of aircraft carriers currently in service with the US Navy. In addition, there appear to be six submarines, and almost 10 destroyers and cruisers.

Seen here in 1960, the Atlantic Reserve Fleet slowly accumulates rust. The two smaller warships on the right are the destroyers USS Monsen and USS Baldwin.

Cruiser USS Dayton (CL-105) at her mooring in Boston, 1960. A year later the cruiser was struck from the naval register and scrapped.

Murphy Army Hospital in Waltham was later converted into the headquarters of Army Corps of Engineers for New England. It was subsequently demolished and replaced with Gann Academy, a private-Jewish high school.

## Otis Air Force Base: The First Line of Defense

F-94 fighters from the 437th Fighter-Interceptor Squadron patrol over the outer Cape in 1953.

Support crews for EC-121 Warning Stars pose with one of their aircraft at an unknown date.

A BOMARC launch station nears completion in July, 1959. (Courtesy Army Corps of Engineers)

A massive BOMARC nuclear cruise-missile rises out of its launch pad. Officially designed to intercept clusters of bombers coming over the ocean, some wonder if the missile was really designed to impact on land in the Soviet Union.

The crew of an EC-121 Warning Star poses in front of their aircraft in the early 1960s. This same aircraft crashed off of Nantucket, 40 minutes after take-off on November 11, 1966, claiming the lives of 19 crewmembers.

North Truro Aircraft Warning Center on the outer Cape, a major Air Defense Command radar outpost. (Courtesy Army Corps of Engineers)

An F-100 Super Sabre from the 101st Fighter Squadron of Massachusetts Air National Guard waits on the tarmac in 1971 as the pilot conducts pre-flight checks.

Seven Coast Guard HU-16E Albatross amphibians wait for their aircrews. The Coast Guard operated the aircraft at Otis between the late ‘60s and the early ‘80s.

An F-106 Delta Dart takes off with full afterburner from a rain-soaked runway at some point in the late ‘70s or early ‘80s.

An F-106 intercepts a Soviet Tu-96 Bear bomber on a reconaissance mission off of Cape Cod on April 15, 1982. The F-106 was retired in 1987, and this aircraft was shot down and destroyed in a test of the hand-held FIM-92 Stinger missile on August 27, 1996.

# Biowarfare

The Morrill Science Center at the University of Massachusetts-Amherst hosted a lab run by brilliant geneticist Curtis B. Thorne between the late 1960s and 1989. A former researcher at the Army germ weapons facility at Fort Detrick, Maryland, Thorne researched the deadly bacteria anthrax to develop a vaccine. However, his research money came mainly from the Army, leading local peace activists to picket his labs in fear that his research would be used to develop bio-weapons for the battlefield.

# Army Natick Labs

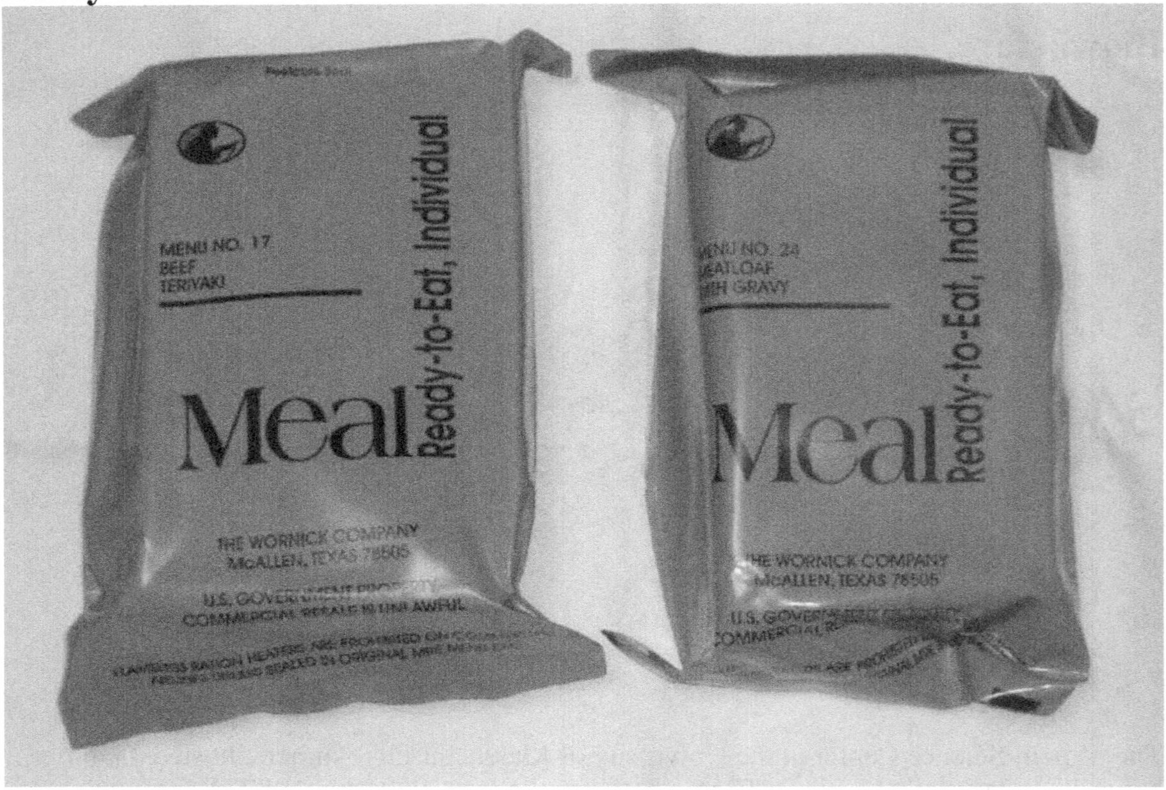


The MRE (Meal Ready to Eat) are dehydrated, irradiated meal packs with long service lives that can be eaten on the homefront or in a combat zone. MREs were developed at Natick to replace cumbersome Meal, Combat, Individual rations in 1981. The MREs shown in the picture (beef teriyaki and meatloaf with gravy) were issued to troops in 2007.

## Chapter IV: The '80s—Back to the Brink

The '80s: it's a decade that conjures images of Farm Aid, the Cosby Show, Star Wars, Ferris Bueller taking the day off, and the kids at Shermer High School writing Principal Verner an essay worth remembering. Internationally, the '80s marked the start of the so-called "Second Cold War." Following the relative peace brought about by the detente in the '70s, the Cold War was threatening to go hot once again. Carter's replacement was actor and septuagenarian, Ronald Reagan, a leader with a firm stance on communism and the Soviet Union.

As president, Reagan rolled out a huge defense budget, totaling more than a trillion dollars annually, to supply the US and its allies with the latest in military technology. The US purchased a new generation of planes, tanks, missiles, and ships, including weapons such as the F-18 fighter, the B-1 bomber, the Abrams tank, the Aegis-class of destroyers, and the Pershing II missile. Reagan's administration faced a nation that was increasingly anti-Soviet, but still hesitant to get involved in foreign conflicts after the bloody quagmire in Vietnam. Instead of military intervention, the US bankrolled a wide range of governments and rebel movements with anti-communist leanings, funding regimes in Central and South America, Africa, and the Middle East. Each year, taxpayers poured as much money as they had spent annually on the Vietnam War into just one of the half-dozen or so major proxy wars that the US was bankrolling, with more than five billion dollars in US aid going to the military government of El Salvador alone.

In New England, the military attempted to 'step up its game,' while major corporations scored substantial defense contracts with the federal government. The late '70s and the early '80s were a low point for many Northeastern cities, and communities throughout the countries. New England was trapped in the grips of the Rust Belt together with the Mid-West, and much of Pennsylvania, New York and New Jersey. Crime rates soared and unemployment skyrocketed. Adding fuel to the fire, a rapid liberalization in attitudes to illegal drugs had taken place since the mid-1960s. 1979 had witnessed the highest levels of illegal drug use in American history, with 25 million people abusing drugs. One in every ten people in the nation was using illegal drugs on a regular basis.

At military bases, low morale, violent crime and drug abuse were rampant, leading military police and officers to crack down as the '80s got underway. Fort Devens remained the largest base of any kind in the region, with a population of nearly 7,000. With many young, restless soldiers on base, in a comparatively sleepy part of Massachusetts, trouble was quick to emerge.

Pvt. Erik Aarhus, a highly trained Green Beret, self-dubbed "Captain Beyond" was convicted on February 8, 1980 of the murder of 22 year old Pvt. Elaine Tyree in the winter of 1979. The young mother of a two year old child had been discovered by police on the floor of her Washington street apartment in Ayer, still dressed in her camouflage fatigues. Aarhus had stabbed her 16 times and slashed her throat, as part of a $500 contract kill arranged by Elaine Tyree's husband, also a soldier. Pvt. William Tyree, the mastermind of the murder was prosecuted separately. Jurors listened to a chilling tape of Tyree giving Erik Aarhus instructions on killing his wife.

"I have the perfect plan...After you do it, take her blood and write 'You're next Tyree' on the TV set or the wall."[134]

In a separate case, two years later, Michael Murra, a soldier from Tacoma, Washington was re-indicted for a rape, committed in broad daylight, on September 1, 1977. Four soldiers had abducted a 16 year old girl off the street within sight of the Ayer police station. The girl had managed to identify Murra, who was sentenced to 10 years in Walpole maximum security prison. However, the other three assailants escaped. After little more than nine months behind bars, Murra was released due to a jury technicality. But the courts were not done with Murra yet. He was reindicted in June, 1981, resulting in a hung jury, with a final re-indictment planned in 1982.[135] Murra was ultimately released with time served.

The ultimate case of off-base hijinks involved Ft. Devens came in 1983, when Sgt. Marc McDonnell, a 27-year old demolition expert was indicted for sending an extortion letter through the mail. The letter was addressed to Kenneth Olsen, president of Digital Equipment Corporation—then one of the nation's leading computer firms in Massachusetts. Claiming to represent a terrorist paramilitary group, McDonnell demanded 1.25 million dollars from Olsen in the form of an 'interest free loan.' To drive home the point, he set off two small bombs in Lincoln in June, 1982, damaging utility poles within a mile of Olsen's house.

When the FBI arrest McDonnell in Germany, he admitted that the extortion was designed to net enough money that his wife could quit work. After his capture on June 23, 1983, McDonnell led investigators to a cache of explosives that he had buried outside of Ft. Devens. When pressed for his reasons, the indicted soldier claimed that, "At no time was physical contact or action contemplated against Mr. Olsen." He had spotted the Digital CEO in Forbes list of the 400 richest people in America. "I narrowed the possible candidates down to seven. From those, only Mr. Olsen lives a near-normal lifestyle in a rural setting. This was a requirement because I felt the operation required a demonstration of capability which could only be done in a woodsy type area where nobody could be injured."[136]

At Pease AFB, 33 security guards were rounded up and arrested in August, 1980. Air Force officials initially stated that the guards had been smoking marijuana while guarding the base's nuclear arsenal. However, they quickly removed references to nuclear weapons to avoid public alarm.

Lt. Nancy Barnett, the base's public affairs officer told reporters, "SAC is very strict on the use of any kind of illegal substance...The other commands are not as strict as SAC. SAC is less tolerant because of the nature of its mission. If a SAC man is on duty, he's supposed to be on alert at all times."[137]

Trouble was not limited to on base crimes. According to the *Boston Globe*, Civil Defense efforts were in a shambles by the early 1980s. Boston's extensive network of air raid sirens, assembled during the early years of the Cold War had not worked since the massive blizzard of 1978. Boston's deputy Civil Defense director, Frank Roumacher told reporters, "The best we could do is have police, fire and other public safety vehicles cruise the streets with their sirens and public address systems on... We have no money to do the job (repair the sirens). There's not a penny in our budget."

"In case of attack we'll just kiss the entire city and its people goodbye... We've got a City Council that thinks Civil Defense is a joke so they won't give us any money to do anything," added Civil Defense Director Harold Good.

The situation was bad everywhere, not just in Boston. In Worcester, a survey in the 1960s had created a list of 460 fallout shelters to protect residents. By the early '80s, many of the shelters on the list had not been serviced in years, and the buildings that housed them had been torn down. In Portland, Maine, air raid sirens had not worked in years. The same was true for Portsmouth, New Hampshire, even though the city was a first line Soviet target, with a major shipyard and one of the nation's largest SAC bomber bases. Officials in Manchester, New Hampshire still had working air raid sirens, but hesitated to test them, worrying that they would incite mass panic in the city.

Associated Press reporters throughout northern New England visited over 100 different fallout shelters, and discovered that almost none of them had working radiation monitoring equipment—in some cases there was no equipment at all. In Hartford, Vermont, none of the local residents cared enough to know the location of the town's four fallout shelters.[138]

In addition to frequent drug abuse, violence, and decay, New England's base's remained trapped in transition. The 1970s witnessed the widespread introduction of women into the military in roles other than nurses and secretaries. The transition to a largely integrated military was not without hiccups.

With the gender integration of most universities in the early 1970s, women were allowed to participate in ROTC programs. In October, 1980, Bonnie Gresham, a Boston University Army ROTC cadet from Boston University was killed when her parachute failed to open at Fort Devens (it was the first fatal parachuting accident at the base in six years.)[139]

In spite of the changes afoot in the military, many service members still hailed from a much different era. Jeremiah Murphy, a reporter with the *Boston Globe* interviewed General George S. Patton IV, the son of famed World War II general George S. Patton, in November, 1980. Patton lived in Hamilton, Massachusetts, although the interview focused on remarks made at the Friends of the Wenham Public Library meeting. In his speech there, Patton spoke out against the anti-war activism of actress Jane Fonda. In Patton's belief, the blood of his soldiers was on Fonda's hands. He finished, "As far as I'm concerned, they can put her up against a wall."

Patton often worked at Fort Devens, after his return from Vietnam. During his time in Vietnam, he had made a name for himself as a firebrand in his father's image, proudly wearing the skull of a Vietnamese guerilla around his neck during a party, and sending out very unusual Christmas cards. They showed a heap of dead, Vietnamese Viet Cong fighters, tagged with the words 'Peace on Earth.'[140]

At the start of the 1980-'81 school year at UMASS-Amherst, the town of Amherst ran into a serious water shortage. Hot weather raised water consumption in the 41 dorms on campus, while a lack of local water supplies led to critically low pressure. Water couldn't reach the upper floors of the massive Southwest dormitory towers. Town officials had failed to warn university administrators about the threat of low water pressure. Although much of the massive campus was made of brick and concrete, the threat of a fire without a reliable water supply to battle the flames could be disastrous.

Flames were a particularly serious hazard in 1980. Hundreds, perhaps even thousands of students smoked cigarettes, and in some cases stronger stuff: marijuana. Without water even a tiny room fire could turn catastrophic. The Army National Guard at Westover stepped in to help, supplying three tankers full of water to battle any flames that cropped up, and to keep remaining students supplied. University administrators ultimately decided to send students home. Thousands left during the first days of the semester aboard 60 buses and 8,000 cars. Students with cars were asked to put paper signs in their windows to indicate what towns they were heading to, so that others might be able to carpool with them.[141]

As the Carter-administration came to a close in the final months of 1980, bases throughout the region witnessed a spike in activity.

Cape Cod was at the center of the action. On September 24, two supersonic F-106 fighters from the 102nd Fighter Interceptor Wing streaked away from Otis for a point over the ocean, 200 nautical miles east of Nantucket. There, they intercepted multiple Soviet Tu-95 Bear bombers on their way south. As the F-106s from Massachusetts handed the Bears over to fighters from Virginia's Langley AFB, and turned for home to refuel, Air Force radar operators onshore tracked the bombers as they headed south. Although the Air Force was not certain where the Tu-95s were headed, it seemed clear that their objective was to reach bases in Cuba where they would refuel to return to the Soviet Union. In addition to building ties with the Cubans, it gave the Soviet pilots a chance to probe America's coastal air defenses. In a war, they would inevitably fly those skies to knock out US military bases, factories and cities.

Major Dick Penney said of the Soviet overflight, "It's not an everyday occurrence, but it's not an unusual occurrence." He added, "They're out there and they're testing US reactions... They test US air space to see if we will do anything if they penetrate. They're long-range training flights for the Russians."[142]

While pilots sought out Soviet aircraft off of Nantucket, ground forces totaling 1,000 civilians, 500 soldiers, and 35,000 reservists quietly kept up operations at Camp Edwards. In 1980, the base hardly looked active. Due to budget cuts in the '70s, the main gate had been boarded up and left unguarded, giving the place an abandoned feel. In fact, Camp Edwards was adding $60 million each year to the local economy.

At the time, the base was one of the few combined service facilities in the country. The Army National Guard alone managed an air unit with 71 pilots and 51 aircraft, an artillery range, railroad, 52 tanks and dozens more armored personnel carriers. The Army was joined by the Air Force, with its remnant Air National Guard operations at Otis, and a $75 million PAVE PAWS radar tracking facility. Camp Edwards also hosted a Coast Guard rescue unit, a government surplus lot, graves for 300,000 veterans at the new National Cemetery, and the 1st Battalion of the 25th Marines.

Activity at the base soared during the summer, when reservists from throughout the Northeast poured in to train. At other times of year, PAVE PAWs construction and the tank refurbishment helped to bring in money and activity at the base. PAVE PAWs was a revolutionary system that is still active in the 2010s on Cape Cod.

The massive radar array assembled on the Cape takes advantage of the natural geography of the region. Cape Cod protrudes out to sea and happens to be in one of the most easterly locations in the continental United States. Therefore, a radar array at Otis would be able to 'look' up and down the east coast from Florida to Nova Scotia. PAVE PAWs was designed to precisely spot anything coming over the water. However, its outstanding capabilities allowed it to issue an early warning of Soviet submarine-launched ballistic missiles rising out of the ocean.

In addition to PAVE PAWs, the Department of Defense also invested in the development of a new infantry grenade launcher at Camp Edwards.

"Our mission is to defend the North American Continent against air attack and we are a prime defense unit. Between our wing here, another at Niag[a]ra Falls, N.Y., and a third at Rome, N.Y., we protect 50 million people," commented Air National Guard Colonel John Olsen.[143]

Unfortunately, not all of the developments at Otis and Camp Edwards were good for local communities. During the construction of PAVE PAWs, many locals voiced concerns about the microwave radiation output from the facility. In fact, the bigger threat may have been the grenade launcher research and development taking place.

In January, 1981, the State Police launched a "purely voluntary" ammunition search in 125 Sandwich homes. A week earlier, a five year old boy in the town had returned home from playing in the woods with an anti-personnel grenade. The boy's mother was initially uncertain about what the object was, and stashed it in her house for days until realizing her mistake. She hastily summoned the State Police bomb squad. The search was led by the State Police working with soldiers from the Army National Guard who combed the woods in search of numerous loose mines and grenades.[144]

The Iranian hostage crisis and the accompanying energy crisis in America carried over into 1981, and nearly brought things to a screeching halt in New England. In January, Massachusetts Governor Edward J. King declared a statewide energy emergency. Workers throughout the state donned sweaters and jackets as thermostats were turned down in factories and offices. The Boston Gas company had lodged a special request with King for the emergency declaration as its natural gas supply ran dangerously low due to cold winter temperatures.

The *Larbi ben M'Hidi,* an Algerian owned liquefied natural gas tanker that delivered most of Metro Boston's gas supplies managed to save the city from the specter of emergency resupply when it sailed on the next tide.

US involvement in the Cold War was not one sided, the Soviet Union and other Warsaw Pact nations were stepping up their game as well. In 1979, the Red Army invaded Afghanistan launching a protracted 10 year battle to instate communist rule. Meanwhile Soviet military advisors, with aid and weapons countered the US across the globe in war zones as varied as Grenada and Angola. The use of other groups as pawns in the larger conflict was nothing new, but the USSR's leadership was changing, becoming more extreme, isolated, and elderly. Most members of the Soviet government who held actual power were older than 70. After Leonid Brezhnev's death in 1982, the Soviet Union had two short lasting leaders—Yuri Andropov and Konstantin Chernenko. Both men were extremely ill and drifted along in hospital beds while running the world's largest country. Andropov, the former head of the KGB had never been to the US and deeply distrusted Reagan.

While undergoing dialysis in a Moscow hospital bed, Andropov's paranoia began to grow. In 1983, the KGB and other Soviet intelligence were busy assembling a web of information that indicated a US "first strike" nuclear attack to wipe out the USSR and its allies. With the world's largest intelligence network, the USSR had plenty of suspicious information to worry over. Watching factors as varied as global prices for medicine, blood, fuel and other commodities, as well as every move made by the US government and military, the picture of a US conspiracy continued to grow.

It would probably be accurate to describe 1983 as the most dangerous year in American history other than 1962 (the year of the Cuban missile crisis). As KGB agents fed the USSR's aging leaders information on the US and NATO, Reagan's administration made a number of moves that heightened tensions. In 1983, Reagan described the Soviet Union as an "evil empire," in a vitriolic speech that frightened the Soviet leadership. On September 1, the USSR gave credence to Reagan's speech when a fighter plane was ordered to shoot down Korean Airlines Flight 007 over the Sea of Okhotsk. Due to a navigational error, the passenger jet had strayed into Soviet airspace and was confused with an American spy plane. When Soviet pilots intercepted the airliner, they found it flying with all fuselage lights on—a course of action that a covert reconnaissance plane would never follow. Although the pilots almost certainly realized that the plane was a civilian airliner, they were ordered to shoot it down when the plane ignored their warning shots (unfortunately, the crew could not see the warning shots, because the fighter plane was not equipped with flare-like tracers). When Flight 007 began to shift to a different altitude, seen as an evasive move, the Soviet pilot was ordered to shoot it down.

The shoot down of flight 007 generated an international outrage. Aside from being an incredibly brazen act, the shoot down cost the lives of a number of Americans. Ironically, one of the men listed among the dead was Congressman Larry McDonald, a politician from Georgia and then president of the ultra-conservative John Birch Society (based in Belmont, Massachusetts). It would seem that America's most outspoken conservative had good grounds for his anti-communist viewpoints.

Autumn, 1983 heightened US-USSR tensions even more. In September, even as fallout from the 007 shoot down spread across the globe, Colonel Stanislav Petrov, an officer subbing in for a friend at a Soviet nuclear early warning center managed to single handedly avert a nuclear war. As the sun set over the US on September 26, an early warning satellite began to register a string of launches of ICBM launches from US Mid-Western bases. Suspecting a malfunction, Petrov continually overrode the computer until the crisis passed. As it turned out, poorly manufactured sensors had picked up infrared light glinting off of high altitude clouds. If not for Petrov's cool in that moment, the Soviet Union would have conducted a massive launch to respond to the US "attack." In spite of his brave actions, Petrov was subsequently suspended, interrogated, and expelled from the military.

On October 25, 1983, even as the Soviet leadership recovered from their attack scare the month before, the US launched Operation Urgent Fury, an invasion of Grenada to rescue American citizens and remove the newly formed communist government. Overlapping all of these other causes for consternation, the US deployed the first of its mobile Pershing II missiles to West Germany in the fall of 1983. The new variety of the Pershing missile was ready to launch within minutes of a warning, and could reach Moscow in eight minutes flight time. The final straw in the 1983 nuclear crisis came in

November when a NATO communications exercise, code-named Able Archer 83, simulated the orders for nuclear escalation that would be given in an actual conflict with the USSR. The USSR believed that the multi-day exercise was a cover for a real life "first strike," and deployed its armies and its submarines in expectation. Fortunately, the order to launch was never given, and hundreds of millions of lives were saved.

Even after Andropov's death tensions would remain high through the '80s, with the US working toward a negotiated drawdown of weapons and an end to the conflict.

On the home front, New England was caught in the cross hairs of the USSR during the crisis years. If the USSR had launched an all-out nuclear war in 1983, or any other year in the '80s, very few New Englanders would have survived. Although the exact details of Soviet targeting are hazy, the most probable targets are easy to infer. Boston, Providence, Worcester and other major cities were almost certainly high on the list of Soviet targets, second only to military installations. With bases spread out across the region, warheads would have hit and heavily damaged almost every New England states except for Vermont (although Vermont would have experienced deadly fallout from a nuclear attack on Plattsburgh AFB, and other major bases spread across upstate New York).

According to FEMA nuclear threat maps published in 1990, most major cities and towns were viewed as probable targets, but in particular centers of industry, military, and state capitals. Except for drifting fallout from airbases in New York, Vermont and New Hampshire were both expected to fare comparatively well in an attack with Burlington,

Montpelier, Concord, and Portsmouth seen as the only targets worth a warhead. Southern New England and Maine would have fared much worse in an attack. Greater Boston, especially the high-tech towns along I-495, and the Air Force base in Bedford were all considered high priority targets, along with New Bedford and Otis AFB.

In Western Massachusetts, Northampton, Springfield, Westover AFB, and Pittsfield were considered high threat. For a time Amherst, Massachusetts would also have topped the list in Massachusetts with plans to build a hardened communications bunker intended to facilitate military communication during a prolonged nuclear war.

Faring almost as poorly as Massachusetts, Rhode Island and Connecticut were both expected to receive a heavy bomb load. Along with Providence, Rhode Island targets included the shuttered but still potentially useful naval station at Quonset Point, and the active naval base in Newport. Connecticut's submarine base in Groton ranked high on the list along with Bradley International Airport, Hartford, New Haven, Stamford, and the other major cities, many involved in producing military aircraft (particularly the Sikorsky and Pratt & Whitney plants).

Maine, the largest of the New England states was also target rich. Along with main coastal towns, northern Maine was expected to receive heavy damage around Limestone, with Loring AFB as a number one priority for Soviet missiles.

This already extremely terrifying picture of nuclear destruction in a densely populated area becomes even more sobering, considering the enormous nuclear fallout from the rest of the country. The New England states were low on the fallout index, and rated as comparatively "safe" compared with Western states like Colorado, Wyoming, Missouri, Nebraska, Montana and the Dakotas—states that would be literally carpet bombed from space due to high concentrations of missile fields.

With Soviet submarines within 200 miles of the shore, there would be no advance warning of an attack. Missiles could hit New England within minutes of launch. If an attack had taken place, millions of New Englanders would have died within the first minutes. For those left, death would be slow and painful, even among the lucky few who somehow managed to find shelter. Adding to the already heavy regional fallout, New England would be blanketed with the accumulated radioactive dust of the entire nation as fallout laden clouds scudded from west to east. For those who did make it through the initial attacks, surviving the fallout would be only the first part of staying alive. They would need to find food, shelter, and clean water, while dealing with looting and lawlessness.

As if nuclear weapons were not enough, the USSR stepped up its efforts to exterminate Americans during the '80s, employing tens of thousands of scientists developing genetically modified biologic weapons on an industrial scale. The USSR successfully weaponized the smallpox virus, even modifying it for extra virulence. According to Ken Alibek, the former chief of the Soviet germ weapons division—Biopreparat—during the '80s, the Soviet Union developed special, refrigerated warheads emplaced above the Arctic circle. Each warhead could be filled smallpox, or any other bioweapon and fired at the US days, or even weeks after a nuclear attack to kill off any survivors. Alibek attests that Boston, along with Seattle and other major cities was targeted for a follow up strike to "clean up," whatever was left over from the nuclear genocide. In an ironic twist, the man who once led one of the most concerted efforts to (potentially) kill Americans, is now part of the nation's line of defense. After the collapse of the Soviet Union, Alibek moved to the US in the 1990s, becoming a consultant for the government, helping to prevent al-Qaeda bio-weapons plots.

Although some details, such as the Soviet bioweapons program were well kept secrets, every American realized that a nuclear war would be the end of life as they knew it. Nuclear war and the anti-nuclear movement became center stage issues during the '80s, particularly in politically active New England. The issue of nuclear war even became a part of popular culture, appearing in everything from songs (take for example Alphaville's 1984 smash hit, *Forever Young*) to movies such as *The Day After* and *War Games*.

*The Day After*, a large budget TV movie set in Lawrence, Kansas portrayed the lead up to and aftermath of an attack based on expert predictions. Widely seen across the nation, *The Day After* sent a clear and terrifying message and reminded Americans of the reality of nuclear war. Ronald Reagan was particularly impacted by the movie, which helped to inspire the 'Star Wars' (officially the Strategic Defense Initiative) missile defense programs. *War Games*, although more campy and youth focused, starring a young Matthew Broderick, still helped to bring attention to the issue of nuclear war for young people. In the film, Broderick's character accidentally sparks a nuclear crisis by hacking into the WOPR missile control computer to play a game 'global thermonuclear war.' The

WOPR begins playing through a "real" nuclear war with attacks on the US. In fact, the first "strike" in the film is Loring AFB in Maine.

1983 was the peak year for anti-nuclear activism. In October, 70 anti-nuclear, environmental, feminist and human rights groups banded together for four days of protest, against the upcoming deployment of the 'Euromissiles' in December. Dubbed the Euromissiles by some clever journalist or activist, these missiles were in-fact American Pershing IIs. Designed as portable, truck-launchable short-range ballistic missiles, the Pershings IIs could hit Moscow in as little as eight minutes. Unbeknownst to most protestors, the missiles were being deployed to counter existing Soviet SS-20 missiles. Following a similar design, the SS-20s spread across the East German landscape could hit London, Paris, Amsterdam and Bonn in even less time.

New England activists called the fall of 1983, 'Hot Autumn,' after the much larger protests going in Britain and West Germany. The different groups planned to converge for a march at Cambridge's Sennott Park. Under the umbrella of the New England Campaign to Stop the Euromissiles, the protestors would urge Senator Edward Kennedy to delay funding for the missiles. The march attracted some high-profile names, including Rev. Jesse Jackson, who was launching his 1984 bid for the presidency, and Representative Ed Markey (a Massachusetts Senator as of 2014). Tadahiko Murata, a Hiroshima survivor invited to speak in Massachusetts, was barred from entering the country as a "threat to national security." After marching to the Tip O'Neill federal building in Boston, the protestors planned to carry on the effort at Pease and the Avco Systems Division factory in Wilmington, where missile components were being assembled.[145]

Even as nuclear war attracted popular and media attention parallel issues relating to the Cold War continued to make the news. During the 1970s and 1980s, growing unrest in the Soviet Caucasus mountain republics was noted amongst the large Armenian-American community in Greater Boston. Armenia had been controlled by the Soviet Union since the 1920s, a state of affairs that created increasing anger amongst Armenian nationalists. In addition to working covertly against the Soviet Union, Armenian nationalists were intent on punishing Turkey for the Armenian genocide that had taken place during World War I (the Turkish government had never officially recognized or apologized for the genocide, and denied that it had ever happened).

Tapping into the frustration of Armenian nationalists and the better funding they could receive in the US, Armenian terrorist groups began recruiting secretly in New England. In the late 1970s, members of one group, the Armenian Secret Army for the Liberation of Armenia set up a terrorist training camp in Franklin, Massachusetts, a suburb of Boston along I-495. The group used an Armenian day camp—Camp Haiastan—to train recruits in hand to hand combat and submachine guns.

Armenian terrorist activity in the Boston area reached a peak in 1982. In March, terrorists believed to be associated with a similar group, the Justice Commandos for Armenian Genocide, bombed a store in Cambridge owned by the Turkish consul, Orhan Gunduz. The bombing was followed in May with another attack on the consul in which

Gunduz was shot and killed at the wheel of his car outside of a doughnut shop in Boston. In more than 30 years since, none of the perpetrators have been caught.

After surviving the regional wave of BRAC in the late '70s, Loring AFB was again becoming a vital part of US defense in the '80s. In the 30 years that had elapsed since the base's opening, the property had been expanded to 14,300 acres, with a second massive runway added in 1985. Loring did not suffer from any space constraints, with tarmac covering 1.1 million square yards, with enough space to accommodate most of SAC's aircraft, and the largest weapons storage capacity of any US airbase. The base had a ready reserve of aviation fuel, totaling more than nine million gallons. Compared with the nearest major SAC base—Plattsburgh AFB in upstate New York—Loring ranked far ahead.

The area around Loring was empty enough to allow the Air Force to open a CERT (Conventional Enhanced Release Training) center. In civilian speak, the CERT was a major bomb range where B-52 crews could drop "dud" bombs. One hundred sixty five miles from Loring was another training facility, the Low Level course, where flight crews could practice flying their bombers at extremely low altitudes to simulate radar evasion tactics and low level-bombing. With improvements in Soviet radar during the '70s, the traditionally high-altitude Stratofortresses had to be adapted to extremely low-level missions, as little as 300 feet above the treetops. The greatest drawback to Loring's location was its distance from Nevada, where annual training missions called for crews to drop live bombs on empty desert test ranges.

International affairs had a funny habit of coming knocking in New England in the '80s, although not to the same Earth-shattering extent as the twin 1970s fuel crises. In July, 1981, the Air Force placed six new F-16 fighters under armed guard at Pease. The planes had arrived from a General Dynamics factory in Texas on July 17 and were still there July 28. Painted in desert colors, the aircraft were awaiting the Star of David roundel of the Israeli Air Force. Instead, the Department of State stopped the shipment, following Israeli airstrikes on Beirut, the Lebanese capital, where they weighed in on a fierce civil war. Little more than a month earlier, the Israelis had struck a blow against Saddam Hussein's Iraq, destroying the nuclear reactor that the Iraqi dictator was using to produce weapons-grade uranium.

The planes were on order for Iran when the Revolution swept through in 1979, leaving General Dynamics to search for new purchasers. The *Boston Globe* interviewed Paul Brockelman, a UNH philosophy professor and supporter of Israel, who admitted that he felt confused about US plans to send weapons to the small Middle Eastern country. "I don't trust the PLO and I don't think we should compromise our support for Israel, but with what seems to be virtually indiscriminate bombing runs by Israel, I actually don't know what we should do..."[146]

New Hampshire Quakers were quick to weigh in on the issue, arriving at Pease for a peace vigil. "In a sensitive region like the Middle East, the proliferation of sophisticated weapons systems provides security and stability for no one," explained Arnie Alpert, a spokesman for the group.[147]

As part of the Reagan-administration effort to revitalize the Armed Forces, Westover hosted a huge exercise reminiscent of the drills of the '50s and '60s, at the end of August, 1981. Operation Condor Redoubt '81 came right on the heels of a small scale air battle between the US and Muammar Qaddafi's Libyan Air Force over the Gulf of Sidra. Unlike the high-altitude combat that Americans were seeing in North Africa, the exercise simulated ground-level skirmishing typical of a Western European battlefield, in Germany or Italy.

The exercise extended far beyond the fenced enclosure of Westover to bases in Newfoundland, the Azores, Britain and Panama, drawing on nearly 3,000 reservists and 59,000 personnel in total. Complete with 'chemical attacks,' helicopter airlifts and fast-maneuvering supersonic fighters, the exercise imagined a fictional country stretching from Quebec to the Carolinas that had seized an allied airliner and accused its passengers of spying.

With its long runways, Westover was at the center of the action, helping an 'ally' headquartered out of Goose Bay airport in Labrador to battle the malevolent, fictitious superpower. Through the exercise, mock saboteurs tested base defenses, attempting to scale fences and break into hangars, while F-4 Phantoms and A-10 attack planes screamed through the sky on sorties and escorted C-130s loaded with troops.[148]

The return to large scale exercises in the '80s was a big shift. Few exercises were as impressive as Exercise REFORGER (Return of Forces to Germany). Simulating a sudden Soviet invasion, REFORGER had begun in 1969, but it hit its stride in the '80s. In fact, the largest American military exercise since World War II was held in 1988, with the arrival of 125,000 troops in West Germany. REFORGER was scaled back and ultimately cancelled in 1993.

In November, 1981, Aeroflot, the Soviet airline was temporarily banned from the skies, after several overflights of "sensitive areas" in southern New England. As a state-run airline, closely tied to the Soviet Air Force and KGB, it seemed that the pilots had been instructed to fly over Groton, perhaps to get some high-altitude photographs of the first Trident submarines stationed there. Alternatively, the overflights may have targeted the new PAVE PAWS array on Cape Cod. This type of furtive, Soviet espionage was a bit more discreet than the spying a year earlier. A Soviet reconnaissance ship hovered near Groton on Long Island Sound, awaiting the activation of the first Trident missile submarine.[149]

The deadly suicide bombing at the Beirut Airport on October 23, 1983, which claimed the lives of nearly 230 US Marines, also claimed nine victims from Massachusetts. One of them, 22-year old, Sgt. Steven Lariviere, was buried in Chicopee. "Some people ask why we call ourselves a nation under God... Just as God has given us a sacred trust with our faith, so, separate but parallel God has given us a sacred trust with our peace," Lariviere's sister wrote on a prayer card.

Chicopee Bishop McGuire, presiding over the funeral said, "When an act of terrorism abruptly ends the lives of innocent young men, the temptation is to strike back, demanding an eye for an eye, a tooth for a tooth. This is not the way to peace."[150]

In the immediate aftermath of the bombing, Leila Laroque, the mother of 19-year old Marine Hubert Laroque, was awaked by a phone call at three o'clock in the morning. The person on the other end of the line identified themselves as a Marine lieutenant, who told her that her son had been badly injured in the attack. Every mother's worst nightmare turned out to be a cruel hoax. Later on that day, Laroque was reassured by a public affairs officer that her son was alright. "[Uniformed Marines will always be sent, even if someone is only injured]. They get an address and they will search and research to find that address. If they have to drive through a field of cow manure, so be it," explained the *real* officer.

In spite of the frequent protests against military activity throughout the early to mid-'80s, the presence of so much equipment and so many highly trained personnel did occasionally prove useful. On March 20, 1985, 66-train cars derailed and caught fire in Greenfield, Massachusetts. Columns of thick, choking black smoke towered into the sky, visible for 20 miles. It was a shock for the sleepy manufacturing town.

What started out as merely an annoyance quickly spiraled out of control. Flames licked at surrounding woodlands and scorched the caulking on a bridge 400 feet away. The clouds of burning octyl alcohol, a petroleum-based solvent, posed a hazard to nearby Greenfield High School and a small Christian school. Three thousand people were evacuated to safety at National Guard armories, while trucks carrying airport firefighters from Westover raced up I-91 to battle the blaze.

One woman told reporters about her harrowing experience with the blaze. As she prepared to take her young son for a walk, the police raced by her house, shouting, "Get out! Get out! There's poison gas coming!" Glancing around, she saw a mushroom cloud of smoke rising into the sky only a quarter-mile away.[151]

By 1985, with Mikhail Gorbachev serving as the new and much more sane leader of the Soviet Union, protestors turned from nuclear Armageddon to US involvements in proxy wars around the world. Even in the '70s, the large-scale US sponsorship of the civil wars in places such as Angola and Mozambique had attracted little attention. Instead, it was the victory of left-wing Sandinista rebels in Nicaragua, in 1979 that kicked off one of the largest protest movements of the '80s.

The poor nations of Central America had spent the '60s and '70s teetering between right and left-wing factions. By the '80s, left-wing groups were in control of Nicaragua, while tiny countries such as El Salvador remained mired in civil war (the US spent almost as much on weapons and supplies for the Salvadoran civil war as it spent on Vietnam).

In June 1985, 51 protestors were arrested at Otis, demonstrating in support of Nicaraguan rebels. "There was no violence, no problems at all. It was very peaceful... We were waiting for them to arrive," said State Police Corporal Ronald Pelletier. The protestors were held in Barnstable for the night, while some were released on 15 dollars bail.[152]

Protests picked up again on August 9, 1985 as anti-nuclear activists marked the 40th anniversary of the Nagasaki atomic attack. The protests centered around the Hanford site in Washington state, where the plutonium for the bomb had been processed. Other demonstrators queued up at Pease, where the unit that had dropped the bomb was then stationed. Two people snuck onto the base to shower ashes on an Air Force statue. They were quickly arrested.[153]

1985 was also a year of revelations about serious pollution at Otis, from contaminants that had built up since the 1960s. Residents in the Ashumet Valley area of Falmouth were urged to use bottled water for cooking and drinking, rather than the usual well-water, in August. Wells in the area had been identified as heavily polluted in 1979 by state officials.

The four contaminants in question were all volatile organic solvents used to clean Air Force jets between 1952 and 1974. Seeping out of sewage lines and leaching from runways, they had formed a slowly spreading plume in the groundwater below, with serious potential consequences for liver and kidney health. Although public wells had been closed, many people continued to draw from contaminated private wells into the mid-'80s. Stretching for 13,000 feet, with a width of nearly 3,500 feet, the plume threatened the Cape's largest aquifer next to and beneath the base.[154]

Even though the Cold War was winding down by the late '80s, neither the Americans nor the Soviets realized it. Base activity remained high throughout the time.

At Fort Devens, the scandals of the '70s and very early '80s continued almost unabated, even though the Army had supposedly cleaned up its act. In July, 1986, a base hospital worker, an intelligence analyst and a dentist were arrested along with two others for selling crack cocaine on-base.[155]

Meanwhile, bad news continued to arrive from Otis. Under pressure from local residents, the state expanded a cancer study across five Cape towns. Preliminary results were not particularly promising. Between 1969 and 1983, rates of lung cancer had shot up 53 percent in Falmouth, with a similar increase for leukemia. Conclusions were hard to draw, but locals were certain that the Otis pollution was behind the change.[156] At the request of the state, the Air Force released information on the microwave outputs from its new PAVE PAWS facility, to contribute to the cancer study.[157]

In spite of the growing controversy about air base pollution, the Massachusetts Air National Guard continued to operate F-106 interceptors out of Otis, until the aircraft's final retirement in 1987. Brand new F-15 fighters were scheduled to arrive for the 101st Fighter Interceptor Squadron in the spring of 1988, a move hailed by Senator Edward Kennedy. Environmental activists were at least somewhat pleased with the aircraft. As Kennedy pointed out, the F-15s only used jet fuel, whereas F-16s carried the deadly toxic fuel hydrazine for backup. The F-15 was seen as an all-around better fighter than the 17, single-engine F-106s, which had just hit their 30th year in service. With its two engines, the F-15 could safely range far out to sea. If one engine was damaged, the other would serve to get the plane and its pilot back to base.[158]

While the Upper Cape Concerned Citizens weighed the potential impact of F-15s, deciding to pursue a cautious optimism, anti-nuclear activists hung their heads at the latest evolution of the Seabrook project. The most expensive power project in New England history, the Seabrook nuclear power plant in New Hampshire had raised the ire of nuclear opponents throughout the region. In September, 1987, Army officials at Fort Devens were asked to consider a plan to use the base as a receiving center for refugees from the communities around Seabrook. If the plant melted down like Three Mile Island or Chernobyl, neighbors would need a safe space to stay and perhaps undergo treatment for radiation sickness. Army officials were wary, given the long distance from the plant. The Dukakis government actively opposed the plan, on the belief that an evacuation near the plant would be impractical.[159]

While environmental concerns remained paramount, the military continued with its training and testing programs. In December, 1988, a pair of Navy jets streaked across a swath of empty woodlands in northern Maine. Soon thereafter, Tomahawk cruise missiles would be dispatched along the same path, flying from a ship offshore to the Navy's Survival School near Saddleback Mountain.

For Christmas, 1988, Senator William Cohen delivered some reassuring, but ultimately inaccurate statements to Maine citizens. At a news conference, Cohen assured Maine residents that both Brunswick and Loring would be safe from BRAC, given their, "valid and validated requirements for national security interests." It seemed to make sense. The Air Force had sunk 300 million dollars into upgrading Loring and was using Brunswick for advanced anti-submarine operations to counter increasingly stealthy Soviet boats.[160]

Such rosy predictions were not limited to Maine. Fort Devens was removed from the list of 86 bases awaiting closure. The Army Intelligence School was set to be moved to Arizona, but it would be replaced by the new Information Services Command, which would bring in an additional 2,000 civilians and three generals. "Fort Devens has survived the cuts," Major General Skaff assured the base.[161]

The last great New England air disaster of the Cold War happened unexpectedly in October, 1989. A KC-135 tanker out of Loring crashed two and a half miles from the Trans-Canada highway in Perth-Andover, New Brunswick.[162]

The four-members of the tanker crew lost in the crash comprised much of the year's regional news related to the Cold War. Although 1989 was a quiet year in New England, militarily, the Cold War was in its final moments. Starting with the opening of Hungary's borders early in 1989 a growing stream of Soviet Bloc citizens began to slip across the Iron Curtain into the West. In almost no time at all, Poland's communist party had yielded power to a democratic government. Throughout the autumn of 1989, protests rocked Czechoslovakia and East Germany. Tens of thousands of East Germans flocked south to Hungary to cross into Austria, until their government banned foreign travel.

East Germany's strict ruling party seemed set to shut down the protests. Members of the feared Stasi secret police were ordered to shoot and kill protestors at a massive demonstration in the city of Leipzig on October 10, 1989. Instead, after 30 years of oppressing the East German people, the Stasi chose not to act. A European equivalent of the Tiananmen Square massacre was averted. On November 9, East Germany officially opened its borders with the west for the first time since 1961. Crowds swarmed over the Berlin Wall, tearing it down piece by piece with heavy machinery and hammers.

The fall of the Berlin Wall became the ultimate symbol of the end of the Cold War, but similar events swept through Eastern Europe, and soon thereafter through communist countries in Africa. At any moment, General Secretary Gorbachev could have ordered the Soviet army to intervene with tanks and troops, but instead, the Soviets surrendered with hardly a whimper.

With the era of superpower conflict coming to an end, it was time for New England to come in from the cold...

# Newport Naval Station

MV Mighty Servant transports *USS Oliver Hazard Perry* to Newport in 1988, for repairs.

## Gerald Bull & the Super Guns

Gerald Bull at McGill University's Space Research Institute in Montreal, 1964. After a promising career as a researcher in Canada, Bull's funding was cut and he set up a company known as Space Research Corporation in North Troy, Vermont. There, Bull and his team developed highly accurate artillery at a 6,000 acre compound spanning the border with Quebec. SRC developed the GC-45 howitzer (below), which surpassed US artillery ranges and sold it to embargoed nations like Iraq, Iran, Israel and South Africa. Bull, a famed artillery consultant was ultimately assasinated by Israel's Mossad intelligence agency in 1990 for attempting to build a massive "supergun" for Saddam Hussein that could strike Israel.

# New Threats: New England in the War on Terrorism

Two Vermont Air National Guard F-16s from the 134th Fighter Squadron take to the skies. Since the '90s, Burlington International Airport and Barnes Airport in Westfield, Massachusetts have become the eastern most air defense bases in the continental US.

A 104th Fighter Wing F-15 takes off from Barnes ANGB in Westfield, Massachusetts.

A C-5 Galaxy from Westover ARB flies over Gillette Stadium in Foxboro on November 11, 2012, in honor of Veterans Day.

102nd Security Forces form defensive perimeters at the Mass. Military Reservation in 2013

## Chapter V: Coming in from the Cold--New England Since the End of the Cold War

The Cold War ended symbolically with the fall of the Berlin Wall, at the end of 1989. In reality, it decayed slowly and continued on in some form until 1995. East Germany, Czechoslovakia, Bulgaria, Bulgaria, and Romania moved to democratic rule in 1990. The two Germanies were re-united in October, 1990.

The Soviet Union remained a powerful force, which still had the potential to challenge the US. But on home soil, the Communist Party was struggling to remain in control. Strong independence movements in the Baltic and Caucasus states were eager to move away from the USSR. At the end of August, 1991, a group of 'hard-liners,' committed to the Soviet Union as it had once been, arrested Gorbachev at his dacha. Their plan called for re-launching the Cold War, and reinstituting communist rule in Eastern Europe.

Instead, ordinary Soviet citizens rose up in opposition to the coup, led by Boris Yeltsin (who subsequently became the first president of the Russian Federation). The coup ended swiftly and the Soviet Union's final execution date was set. The USSR would officially become the Russian Federation, Ukraine, Belarus, Estonia, Latvia, Lithuania, Armenia, Georgia, Azerbaijian, Kazakhstan, Kyrgyzstan, Tajikistan, Uzbekistan and Turkmenistan at the end of December, 1991. Ironically, the coup forestalled a policy change that Gorbachev was working to enact that would have transformed the USSR into a communist version of the European Union—complete with a powerful military, but with greater freedom for its unruly republics.

The final communist states in Europe and Africa, including Yugoslavia, Albania, Cambodia, Mongolia, Afghanistan and Ethiopia disintegrated in 1992 and '93. The last attempt as resurgence by the communists, anywhere in the world, came in the summer of 1994, when a rebel group briefly re-created the communist country of South Yemen.

For the US, the shift away from Cold War mindsets was slow to take place. Soviet troops remained on the ground in East Germany and the Baltic states until the autumn of 1994. The collapse of the Soviet Union left thousands of nuclear warheads lying around unsupervised in the post-Soviet republics, and civil wars were quick to break out across Eastern Europe and Central Asia. The Russian Federation was merely a shadow of the old USSR, but it still had ample fire power to cause serious devastation. In one confused instance in 1995, the Russians nearly launched their nuclear arsenal against the US, after a research rocket from Norway was misidentified as an American Trident missile.

While the Cold War was still officially in progress, Saddam Hussein's Iraq launched an invasion of the small oil-rich emirate of Kuwait. The annexation of Kuwait in August, 1990 elicited an immediate American response. US forces were deployed to protect Saudi Arabia from aggression as part of Operation Desert Shield.

The summer of 1990 had been a relatively quiet and happy one at New England bases. Otis had hosted the Beach Boys for a large concert. The "eternal surfers," arrived on July 4 to play for a crowd of 100,000.

When Desert Shield got underway, Westover was immediately involved. The base's 16 C-5s were pressed into service to haul tanks and troops to Saudi Arabia. "My crews are trained day-in and day-out on this stuff... It gives you a desire to test your skills in combat. They're not reticent about volunteering," commented Lt. Colonel Jim Gallin. A Digital Corporation executive by day, Gallin enjoyed the community spirit at Westover, noting that his own lawyer was part of the squadron. C-5 crews had recently returned from the invasion of Panama in December, 1989, and were geared up for war.

By the end of the month, a large-scale call up of reservists and National Guard members was taking place nationwide. Scores of silver haired, slightly portly Guardsmen assembled for the first time in years at their respective bases. Some had been activated short-term for the Blizzard of '78 and 1985's Hurricane Gloria. But those had been brief activations for natural disasters. David Gavigan, a former high school teacher from Fall River led the operation in New England as the assistant adjutant general of the Massachusetts National Guard. At the age of 23, in 1961, Gavigan had been called up with his students for the Berlin Crisis, but he admitted that it had been years since his unit had activated for war.

Unfortunately for defense contractors, Iraq was not much of a naval power. Therefore, there was no reason to slow or halt cuts to submarine production at General Dynamics-Electric Boat. One thousand workers were laid off at Groton and Quonset Point. A General Dynamics spokesman admitted, "It tells us that we are looking at a considerably smaller amount of business than earlier projections." The corporate term of the day was 'rightsizing.' Over the coming years, New England contractors would witness a great deal of rightsizing.

As the activation continued throughout the fall, many members of the modern military faced unexpected challenges. Major Hickey, a commander in the Air Force's Nursing Corps at Westover was faced with a sudden dilemma when she was summoned to Saudi Arabia. Hickey told reporters about her growing anxiety for her two year old daughter, who would have to be placed in full-time daycare if her husband's Coast Guard commitments called him to the Gulf as well. The couple raced to figure out a will and custody for their child if the Kuwait crisis turned bloody. In the lead-up to her deployment, Major Hickey enjoyed final moments with her daughter, while working through a mountain of papers on her kitchen table. None of the information was very encouraging, as she read about cholera, mustard gas, and daytime high temperatures above 140 degrees. "I don't like it; let's face it," Hickey said, adding, "I'm proud to do my part...you have to roll with the punches. [My daughter's entire comprehension of the Kuwait crisis is] 'Mommy's airplanes.' I want to dance at my daugher's wedding...I don't want my daughter to be an orphan."

By December, as the chance of outright war drew ever closer. Ten thousand protestors converged on Boston Common aiming to stop a conflict. Daniel Ellsberg, the famed anti-Vietnam activist told the crowd, "We have 45 days to stop this war. No president has ever given us that much warning before...We love [our] troops...We're doing our job to bring them back safely." Michael Klare, a Hampshire College peace and world security studies professor announced that, "I think war is likely because Bush has ruled out any other possibility . . . we have to be prepared for a terrifying spectacle." Klare and many others like him believed that a war with Iraq would quickly swell to the size and casualties of Vietnam, and perhaps even World War II, with anywhere between 50,000 and 400,000 dead. In light of Iraq's liberal use of chemical weapons in its war with Iran, such predictions didn't seem entirely unfounded. Not far from the Hampshire campus, 36 people were arrested trying to block Westover's gates.

However, Klare could not have been further from the truth. Efforts to stop the war failed and at the end of January, 1991, hostilities began. A two week air campaign humbled Iraq from the sky, and allowed for a brisk, four-day invasion, which quickly swept the Iraqi's from Kuwait with little more than 100 Coalition combat deaths. It was a stunning victory for the US.

Throughout the lead-up to war and the conflict itself, New England bases like Loring and Westover finally put some part of their Cold War mission to the test. As some of the easternmost airbases in the country, they were perfectly positioned to transport troops, tanks and supplies to the Gulf.

When the conflict started, New England bases were placed on high alert, in the event of a covert Iraqi counterattack plotted by its extensive intelligence services. Random vehicle searches and careful identity checks went into effect at bases as large as Groton and as small as the Air Force radar station in New Boston, New Hampshire.

While service members like Major Hickey, and others like her were deployed to the Gulf War, hundreds of grandparents were essentially drafted into a childcare role. Reporters from the Boston Globe met with Alan and Barbara Goldsmith in Cummington, who had taken on the task of caring for their grandchildren, age 10 and 12. With a

Department of Defense policy of 'worldwide deployability,' there was no guarantee that married service members would not both be sent to a combat zone.

A Holbrook family caught a flight to San Francisco on a moment's notice to collect their three year and 18 month old grandchildren, as the children's parents deployed overseas. For grandchildren that had met their grandparents only three times before, it was almost as if they were being spirited away to live with strangers. Every grandparent that took on the tough responsibility of caring for their grandchildren during the war shared the same fear—that their child and in-law would not return from the battlefield. "It's a horrific thought," one Milton grandfather admitted, "At our age, we are relatively ill equipped to deal with this, and the kid might be left without the possibilities we took for granted for our children, things like going to college." In spite of the constant worry and hardship, he added, "Both my wife and I were career-oriented when we were young and never really had enough time to spend with our own kids. We see more changes in our grandson every day than we ever noticed in our kids 20 years ago. We just love this little guy to death."

When the Persian Gulf War finally ended at the end of February, 1991, America rejoiced. Returning soldiers were greeted with a ticker-tape parade through the streets of Manhattan, while their Iraqi opponents retreated to lick their wounds (nearly 30,000 Iraqis, mostly in the Army, had been killed in the conflict). Jeffrey Fox, a combat pilot from Swansea had been shot down during air operations over Iraq. He returned to the US at the start of March to the sad news that his squadron-mate, Patrick Olson, had been killed in combat.

With the Cold War over and US forces returned from the Gulf, a new and very hopeful future seemed to be in store for the nation. No longer, it seemed, would America need to worry about nuclear annihilation or tanks crawling across the farmland of West Germany's Fulda Gap. But what did this new reality mean for all of the Cold War institutions that still existed?

Civil Defense seemed ready to go the way of the dodo bird. However, A. David Rodham, the civil defense director for the Weld-administration in Massachusetts disagreed. "Civil defense is still very much alive and well." The problem for local Civil Defense directors was funding. Appropriations were quickly slashed at the town and state level, limiting the number of Civil Defense staff available. Medford's director admitted his fears for the future, "I hope and pray nothing serious happens." As a result of decreased funding, Civil Defense was increasingly forced to rely on equipment from local police, public works and fire services.

Rodham explained, "We sometimes explain our agency as the 'what if' agency." Even though the threat of nuclear war had largely disappeared, Civil Defense still intended to prepare for nuclear accidents, terrorism and severe weather. During the Gulf War, they staged evacuation drills at Logan and Westover in case the Iraqis had decided to unleash bombs or chemical weapons at either location.

In light of the new mission—whatever it was, exactly—many bases that had been expected to stay open were quickly moved back onto the BRAC list. Loring was scheduled to shut down in 1994, now that B-52s were no longer needed to hammer the

Soviet Union. The closure would end up costing northern Maine 3,300 vital military jobs, in addition to 1,300 civilian layoffs.

As final operations got underway at Loring, a dozen workers were exposed to radiation after cutting a hole in the side of an old weapons shed. The workers were decontaminated, but the maze of steel doors they glimpsed in the shed hinted at possible abandoned nuclear waste, or perhaps just a hazardous buildup of radon (the building had been sealed air-tight since the '60s). Its entrance had been bricked up to prevent terrorists using it as a fortified command post in the middle of the weapons storage area. Specialists from Texas were flown in to investigate the source of the radiation.

Outside of active military bases, cuts were expected to go deep. Defense Secretary Cheney (the future vice-president under George W. Bush) called for a reduction of nearly 8,500 Guard and reserve jobs in Massachusetts, as well 2,836 in Connecticut, 1,790 in New Hampshire, 1, 267 in Vermont, 1,790 in Rhode Island, and 1,143 in Maine. Older, pensioned soldiers expressed their sympathy for young service members, with only 10 or 15 years under their belts. They would not receive a pension as a result of the cuts. Following on the heels of Fort Devens' closure, it was a major blow in southern New England.

Back up north, the spring of 1992 saw plummeting house prices in Aroostook County as the Air Force prepared to leave. In the small community of Washburn, prices plunged more than 16 percent.

In a last-ditch effort by the Maine congressional delegation to save the base, the Air Force briefly considered moving its 'Star Wars' missile defense post from North Dakota to Loring. The program had yet to yield any results in spite of the billions already spent on it (the plan called for a massive constellation of orbiting satellites to shoot down enemy missiles). Ironically, a move to Loring required the approval of the new Russian Federation, under the terms of the 1972 Anti-Ballistic Missile Treaty.

In a curious twist, residents of northern Maine counties cast votes overwhelmingly in favor of independent candidate Ross Perot in the 1992 presidential election, who sided with workers at Loring and Kittery.

Further south, Brunswick NAS seemed to have brighter prospects. Even though the emphasis had changed away from Soviet submarines, the Navy was reinventing the base for the 1990s. In addition to pursuing submarines from the new Russian fleet, P-3 Orions from the base would track drug smugglers and the smaller submarine forces of countries like Iran, Libya, North Korea and Cuba. Back at base, the walls of the intelligence office were plastered with photos of Soviet subs, as well as aerial snapshots of Algerian and Libyan craft.

Not long after the end of the Gulf War, military personnel throughout New England adapted to the new reality of small, regional wars. C-5s from Westover roared southeastward toward Somalia, to resupply American forces in a UN-sponsored peacekeeping mission. Somalia had plunged into a bloody civil war in 1991 after the fall of its communist dictatorship. The C-5s would fly to Griffiss AFB in upstate New York to collect supplies from the 10th Mountain Division, and would meet up with KC-135 tankers based out of Pease to refuel. Unlike many previous missions, C-5s carried huge quantities of food with them to feed almost one and a half million starving Somalis. Sadly, the heartfelt US mission to Somali ended in disaster following the 1993, Blackhawk Down incident. US troops pulled out in 1994, leaving Somali in a state of anarchy that persists to the present day.

The cuts that had struck first General Dynamics and then Loring and Fort Devens soon caught up with the Portsmouth Naval Shipyard. While the base would not be officially closed, its staff would be reduced from 6,000 to as little as 1,000. For once it seemed that politics intervened on behalf of a New England. The Clinton-administration was eager to gain the support of Senate Majority Leader George Mitchell, from Maine, who was a staunch supporter of the shipyard. Clinton was also eager to win the 1996 New Hampshire primary. Taken together, these two factors helped to save Portsmouth from the ax of funding cuts.

Cuts in 1993 were harsher still. Twenty-one thousand military and civilian jobs would be cut throughout New England by 1999, and South Weymouth NAS would also be axed. New Hampshire residents near Pease (which had closed in 1991) could commiserate with the folks in Weymouth as force levels prepared to drop 33 percent below 1974 levels (1974 had been the last great period of trimming). For communities, base closures were extremely bad news, hurting businesses and radically altering the tax-base virtually overnight. Since 1991, the area around Pease had lost 4,500 jobs—Fort Devens was poised to lose more than 9,000. The disappearance of children from the services would gut the Ayer school system, leaving them with a network of overbuilt and practically empty schools. The closure of South Weymouth would claim 1,050 jobs in total. Between 1988 and 1993, five of the region's largest military facilities closed down.

In Newington, New Hampshire the school system lost 25 percent of its students, closed four schools and quickly laid off 100 teachers following the closure of Pease. Things looked even worse for Limestone, where 85 percent of the school system was set to evaporate by 1994. According to social service workers, each base closure was accompanied by a rapid increase in local child abuse, neglect and alcoholism, as well as cases of domestic violence. In most cases, the military had no clear plan for redeveloping the bases, leaving it up to local officials to scramble to attract large and small businesses.

Realizing the seriousness of the situation, Governor Weld of Massachusetts and his counterparts in Connecticut and New York appealed to the federal government to slow the process. Weld noted that closing South Weymouth would cost the local economy 56 million dollars. "This is a matter of life and death for the region," Senator Lieberman from Connecticut announced firmly. For Lieberman's state, this was probably true. Between 1991 and 1993, Connecticut lost 10,000 defense jobs. Governor Bruce Sundlun of Rhode Island chimed in, describing the closure of Quonset Point 20 years earlier. "[It] nearly devastated our economy, and came close to leaving much of Newport County a ghost town." Unfortunately, the best lobbying efforts of the three governors failed to stop the closures.

In a surprise move, the Russian fleet sailed into Boston harbor for the first time since 1975, at the start of July, 1993. The arrival of the Russians signaled that the Cold War had truly ended. The US and Russia held a six-hour naval exercise off of Cape Cod before proceeding into port. US Navy officer, Captain Gallant, described the event as, "not only getting to meet your counterpart and establish friendships...[but also] getting the Russians into Boston and spending time [together]." A Russian naval band performed at City Hall while Russian sailors jockeyed with US sailors and police officers, in games of hockey and softball respectively. A huge block party shut down Lansdowne Street, as many Russian sailors got their first taste of American life.

While base closings continued and US sailors partied with their Russian counterparts, trouble was brewing in Eastern Europe. The implosion of Yugoslavia in 1992 led to a state of civil war in most of its member republics. Bosnia stood out for its violence, as Bosnian Serbs attacked Bosnian Muslims in a battle for the future of their ethnically and religiously divided country. Serb fighters surrounded the Bosnian capital of Sarajevo (which had hosted the 1984 Winter Olympics) and proceeded to lay siege to the city for months, pounding it with artillery shells, and slaughtering civilians—and children—with sniper fire.

By 1995, NATO was ready to get involved, marking their first ever combat mission with air strikes on Serb forces throughout Bosnia to 'level the playing field.' Throughout the Yugoslav wars, P-3 Orions from Brunswick flew to Italy, where they operated out of forward bases to enforce an embargo on the Federal Republic of

Yugoslavia (the Serb-majority remnant of the old Yugoslavia) which funded and supported ethnic Serbs on the ground in Bosnia. At other points in the mid-1990s, the Orions responded to Haitian boat people fleeing political unrest and carried out reconnaissance over Rwanda during the 1994 genocide. One flyer described the change from old ways of operating, "It's a cultural change for us... It's different, but it is just as exciting."

On May 4, 1994, the last KC-135 left Loring on an air refueling mission. As Aroostook looked toward a bleak future of 25 percent unemployment, the federal government offered a small concession. Three hundred Defense Finance and Accounting Services offices would be consolidated at the shuttered base, helping out the Loring Development Authority with an immediate infusion of 22 million dollars a year. At the end of July, 1994, a massive auction sought to dispose of service members unwanted belongings at Loring. The final step in closing the base was shipping radioactive soil to Utah on 20 rail cars, at a cost of four million dollars. Late in the year, a team from the Department of Justice visited the base to find out if it would be suitable for a federal prison.

Meanwhile, far to the south, General Dynamics was in more pain than ever. A visit to Quonset Point or Groton might not have shown much out of the ordinary. Work crews labored 10 hours a day, six days a week, catching up on a three year backlog of Navy orders. With six submarines in the works, the real question was whether or not the two Seawolf submarines would go through. The ultra-stealthy submarine envisioned as a replacement for the Los Angeles-class attack sub was viewed as wasteful spending by many Republicans, such as Senator John McCain. Instead, they favored using the money from Seawolf to fund new B-2 stealth bombers, which could take out targets in Iraq, Libya, Iran, or North Korea. Like the Portsmouth Naval Shipyard, Electric Boat had been rescued by a 1992 Clinton campaign promise to complete the third Seawolf-class boat (part of a bid to win over Connecticut voters).

Administrators in the government and at Electric Boat had a hard time abandoning Cold War mindsets in favor of the new reality; one in which submarines launched Tomahawk cruise missiles at targets in Iraq, unseen and virtually undetectable. One analyst summed it up, "The problem with the submarine community is they are still hunting for Red October... In the absence of any other rationale for submarines, they are hoping that the bear is going to wake up. It is as though they are stuck in the summer of 1989." Another commentator added, "The fact that the Cold War is over does not mean that our interests and Russia's will always be in line. They build a good sub... The quality of our ships is influenced by that high end threat..."

The closure of South Weymouth NAS in 1997 marked the last large downsizing of the 1990s. In 1999, the US military hit their lowest force levels since before World War II. It seemed that conflict was winding down, even in light of the 1999 US bombing campaign against the Federal Republic of Yugoslavia. Some people may have incorrectly assumed that the cuts would continue, as units of every size contracted, and packed into a decreasing slate of installations. The future, it seemed, held nothing but small regional interventions and peacekeeping missions, like Somalia, Haiti, and the Balkans. Of course, history had other plans.

On the evening of September 10, 2001, a man named Mohammed Atta stayed at the Comfort Inn in South Portland, Maine with another man named Abdulazziz al-Omari. From a location near the motel, Atta called Khalid Sheikh Mohammed, the mastermind of the plot that he had joined. Mohammed gave him a coded signal to proceed.

Early the following morning, the two men boarded a Colgan Air flight from Portland to Logan, where they joined eight other people and boarded American Airlines Flight 11 and United Airlines Flight 175, both en route to Los Angeles. At 8:14 in the morning, Atta and his four co-conspirators leapt out of their seats on Flight 11, and used box cutters to stab flight attendants Barbara Arestegui and Karen Martin. They also slashed the throat of Daniel Lewin, an American-Israeli who had once served with the Israel Defense Forces. By 8:37, an accidental transmission from the cockpit to ground control informed Logan that Atta had hijacked the plane. NORAD was immediately alerted. NORAD quickly routed an emergency alert to F-15s at Otis. The base took a few minutes to respond, but soon scrambled two fighters, flying with full afterburner.

Over Manhattan, at 8:46 Atta completed his final turns and crashed the aircraft into the North Tower of the World Trade Center, at the same moment that Otis received the order (the F-15s didn't make it into the sky until 8:53).

Flight 175 was hijacked around 8:42, narrowly avoided hitting a Delta Air Lines flight, and struck the South Tower of the World Trade Center at 9:03 in the morning.

The terrorist attacks in New York, Washington, and Shankesville on September 11, 2001, kicked off a new era—the so-called War on Terrorism. Billed as a kind of new global, low-level conflict, typical of the 21st century, the War on Terrorism was itself rooted in the Cold War.

Following decades of European imperialism, Israeli success, and Soviet-backed Arab socialist rule, the Middle East was ready for a revolution. The first generation of modern Islamist terrorists picked up the centuries old tradition of jihad in the 1970s; take for instance the Palestinian and Iranian extremists of the era.

Surprisingly, though, it was the Soviet invasion of Afghanistan that gave the Islamists teeth. With US backing, a huge force of mujahedeen warriors armed with AK-47s and shoulder-launched Stinger missiles managed to turn the tide and evict the Soviets from Afghanistan in 1988. The communist government that they had installed held on until 1992 in Kabul and a few other major cities, before the Taliban and their enemies, the Northern Alliance, broke Afghanistan in two.

During the same time, the Soviet Bloc collapsed, and in its final moments the Soviets neglected to protest against the Persian Gulf War. Osama bin Laden had been involved in the US-backed mujahedeen actions of the '80s, but he quickly turned his back on the US. Al-Qaeda took shape in 1987 and began launching attacks in 1993, with the long range goal of creating a global Islamic caliphate (the caliphate was a type of large Islamic state that had once stretched from Afghanistan to Spain; the last official caliph was the Ottoman sultan in 1924). Al-Qaeda's short-term goal was to punish the infidel

Americans for putting troops on the holy sands of Saudi Arabia, too close, it seemed to the holy cities of Mecca and Medina.

Al-Qaeda and groups like it enjoyed tremendous success throughout the 1990s. They destroyed the Khobar Towers apartment block in Saudi Arabia, full of US Air Force personnel, as well as US embassies. Meanwhile, the allied Taliban movement in Afghanistan set up extraordinarily strict sharia law, completed with public executions in stadiums. They also afforded al-Qaeda a place to train its operatives. But the coup de grace did not come until September 11, 2001. The US responded quickly to the attacks, launching its invasion of Afghanistan in October, 2001. Together with the Northern Alliance, the US swept quickly into Kabul and the rest of the country. Unfortunately, in a rarely publicized instance of international betrayal, Pakistan's intelligence service launched the Kunduz airlift. Thousands of Taliban fighters were carried to safety in Pakistani planes, allowing them to continue the war. Thirteen years later, the US and its NATO allies are still committed to Afghanistan, although troop numbers have been vastly scaled back since the main withdrawal in December, 2014

On the home front, the specter of the 9/11 attacks brought out some very ugly behavior, even in traditionally tolerant New England. In one instance, on September 17, 2001, arsonists destroyed a Plymouth pizza shop owned by Iraqi native Salah El-Ehmeada. Standing amidst the scorched ruins of his shop, El-Ehmeada told reporters, "This is un-American. America was not built on this." El-Ehmeada admitted that he had been as shocked by the 9/11 attacks as any other American would be. "I felt like a guy who comes from a good family, a great family...and he wakes up one day and his brother is a serial killer on TV."

The daughter of one of El-Ehmeada's regular customers summed it up best. "He didn't do anything. He was making pizza."

In the immediate aftermath of 9/11, the US was shocked again by a series of mailings, containing deadly anthrax spores, which claimed the lives of five people and costs hundreds of millions to clean-up. Although subsequent evidence seems to indicate that the perpetrator was a disgruntled US Army scientist, many at the time didn't know what to believe and assumed that Islamists (perhaps with Iraqi backing) were behind the attacks. Anthrax is comparatively easy to produce and if high-level sponsors were involved, it was possible that terrorists could roll out smallpox.

Overnight, National Guard troops were deployed outside of the offices of Acambis PLC in Cambridge. The small biotech firm had hardly stood out before 9/11, but now they had become more significant to the nation's defense than almost any other company. Long before 9/11, Acambis had acquired a special license, making it the world's only authorized producer of the smallpox vaccine. No samples of the vaccine had been produced since the threat of Soviet biowarfare in the '80s. Now, the US government was asking Acambis to churn out 40 million doses by 2004, from its labs in Canton, at a total cost of 343 million dollars. Naturally, terrorists might figure this out, and presage a smallpox attack with a truck bomb assault on the unassuming Cambridge company.

In a similar vein, strict security measures went into effect throughout New England. National Guard troops took up posts at Logan, and stationed armored vehicles near the Wachusett and Quabbin Reservoirs to forestall terrorist attacks on the Boston water supply system.

The 9/11 attacks hit home in a very different way for a family in Cambridge—none other than bin Laden family. Abdullah Mohammed bin Laden, a graduate of Harvard Law School, had encountered a sobbing barista at Starbucks the morning of the attack, mumbling something about an attack in New York. Seizing his latte and thrusting a few dollars across the counter, bin Laden raced home to his apartment, sharing in the nation's grief at the tragedy. Soon thereafter, he discovered that the mastermind of the attack was his half-brother Osama bin Laden. Abdullah was one of 54 children that bin Laden patriarch Mohammed Bin-Awad bin Ladin had had with his different wives. The wealthy bin Laden family had helped the Saudi government to build up Mecca and even assisted the US in building bases for the Gulf War; and yet, one of its 54 members (disowned in 1994) had plotted the most devastating terrorist attack in history.

Abdullah explained his own ties to the Boston area. He had first visited the area in 1990, and was immediately entranced. "I was fascinated by the city, by its charm. I felt it was the best of both worlds - America and Europe... I think, `This is the place. I shouldn't go anywhere else.'"

Two years later, the US launched its controversial invasion of Iraq, removing Saddam Hussein and kicking off an ongoing insurgency (which now threatens to topple the Iraqi government). Thousands of New Englanders across all service branches participated in the two wars, and dozens were ultimately killed, particularly in Iraq.

The surge of activity around the Iraq War proved to be a boon for New England contractors. In 2002, contracts were issued for a stealth destroyer (the USS Zumwalt launched in 2013) to combat enemies such as Iraq and Iran. The 'DD-X' project would help to create 200 jobs in Kittery at the Bath Iron Works.

Much of the work in the region focused on bio-security and computer systems. BAE Systems in New Hampshire bought out a Lockheed Martin division in 2000, and set about designing new military computers, together with the 350 new employees working on shipboard electronic systems for Raytheon in Portsmouth, Rhode Island. Raytheon was also heavily involved in creating guidance and optical systems for the Global Hawk surveillance drone and the spy satellites that made the War on Terrorism possible.

At the height of the two wars, nearly 400,000 US troops were deployed in the war zones, in addition to tens of thousands of others in Kuwait, Turkey, Britain, South Korea, Japan, and Germany. Although the conflicts were smaller than Vietnam, they cost American taxpayers much more, between high-priced fuel and extremely expensive equipment. Fortunately, some government contracts did benefit New England companies.

Defense contracts of all kinds poured into providers throughout the region. In one relatively small-scale instance, the British-firm Qinetiq developed and tested body armor for troops and for Stryker armored vehicles out of its factory in Franklin, Massachusetts to aid the Iraq War. In Maine, Bath Iron Works has continued its production of ships throughout the War on Terrorism, unabated. The most recent victory for General Dynamics and Bath Iron Works was the launching of the USS Zumwalt in 2013. The first 'stealth destroyer,' the somewhat bizarre looking vessel is the end result of the DD-X program, which, like the new F-22 fighter is intended to absorb or simply not reflect radar signals, limiting the chances of detection by an enemy force. Similarly, the General Dynamics Electric Boat division has managed to escape from the mire of 1990s budget cuts.

In the mid- to late '90s, it had seemed as if Electric Boat might go belly up any day. Today, in the 2010s, production is soaring and Electric Boat is hiring scores of new engineers and tradespeople for its operations at Quonset Point and Groton. The expensive Seawolf project that had nearly shut down Electric Boat in the 1990s was cancelled, after only three submarines were produced. The project was given the F-16 treatment (F-15 fighters are notoriously expensive due to electronics and twin engines, and cheaper F-16s are often substituted in their place) replaced with the Virginia-class submarine program. A less expensive alternative to Seawolf, the Virginia-class submarines are practically are keeping Groton and its sister station at Norfolk booming. So far, 10 of the submarines have been launched, with 20 more under commission through the early 2030s. The submarines are expected to remain in service until the 2070s, helping to keep Electric Boat up and running through repairs and overhauls. Although billed as the remedy for growing undersea threats around the world, the actual purpose of the Virginia-class had less to do with defense and more to do with politics. The contracts help to satisfy a powerful General Dynamics and voters in Connecticut and Rhode Island, and also help to keep shipbuilding expertise alive in the US (privately run shipyards for civilian vessels may be unable to compete with foreign shipyards, and under the terms of free trade agreements can't be protected by the government).

With the winding down of Iraq War and the War in Afghanistan, the US continues to be heavily involved abroad. It participated briefly in the airstrikes on Libya in 2011, aimed at bringing down the Muammar Qaddafi regime (a long-time coming after the skirmishes of the '80s). Rebel casualties from the Libyan Civil War were flown into Logan and transported to Massachusetts General Hospital.

Although New England seems far removed from the war zones of Iraq, Afghanistan, Yemen, and Pakistan, it had some of the most active terrorist activity anywhere in the US. New England is a populous and cosmopolitan place, with liberal attitudes, bustling airports, and close proximity to the halls of power in New York and Washington. It may come as no surprise that terrorist activity has cropped up frequently in the area.

In October, 2009, 27-year old Sudbury pharmacist Tarek Mehanna was arraigned by federal prosecutors for conspiring to assassinate two prominent politicians and shoot-up shopping malls in the Boston area with automatic weapons. The would-be terrorists used code words like "culinary school"—terrorist training camps—and "peanut butter and jelly"—for the ongoing Islamist effort in Somalia.

The group decided to call off their plans after meeting with trained-terrorist Daniel Maldonado in New Hampshire to buy weapons. Fortunately for Massachusetts shoppers, Maldonado could only offer them handguns instead of assault rifles.

Terrorists struck a small but significant blow on April 15, 2013 in the now infamous Boston Marathon bombings. At 2:49 in the afternoon, as the race drew to a close, twin bombs, improvised from pressure cookers exploded along Boylston Street, by the Boston Public Library and the finish line. It was the first time since the assassination of Turkish consul, Orhan Gunduz, in 1982, that terrorists had claimed lives on the ground in New England.

What ensued over the next few days was a police and media circus. Initially, the police had few well-defined suspects. Reeling from the unexpected attack, the Boston Police, together with field agents from the FBI, ATF, and Homeland Security launched an 'all hands on deck' investigation. Meanwhile, social media sites such as Reddit erupted into a digital witch-hunt, complete with inaccurate tips to the police and false accusations against innocent individuals. The online witch-hunt may have spurred the police to release the images of the two suspected bombers too early.

The suspects were identified as Chechen immigrants Dzhokar Tsarnaev, 19, and his old brother Tamerlan. Realizing that they had been found out the brothers went on the run. On the evening of April 18, they held up a 7-Eleven convenience store in Cambridge, murdered an MIT police officer, and went on the run with a carjacked vehicle. Tamerlan was shot and killed soon thereafter while Dzohkar remained on the run.

What followed was one of the largest manhunts in US history. The dragnet came complete with a massive constitutional breach. Federal investigators, Boston and Watertown police, together with SWAT teams and the National Guard turned much of Greater Boston into a ghost town. The hunt quickly moved to Watertown, where road blocks closed off the entire community. Police stormed through neighborhoods, searching homes one by one, armed with automatic weapons. Throughout the day poor coordination led to frequent friendly fire incidents as different police units shot at one another. Amazingly, it wasn't the police, but rather a homeowner that discovered Dzohkar's hiding place.

A trace of blood on the hull of a boat in the backyard led the homeowner to call police. SWAT teams showed up in force and proceeded to open fire on each other. Dzokhar was shot in the throat and hauled from the boat, although he didn't resist and surrendered easily. Following his capture, Dzohkar was moved to a medical facility at Fort Devens to await his trial for terrorism. In an ironic twist, it was the old enemies—the Russians—that came out of the Boston bombings looking like the 'good guys.' Russian intelligence had warned the US about the Tsarnaev brothers long before the attacks.

The attacks pointed out the realities of War on Terrorism on home soil, and the need for police forces to coordinate more effectively. The attacks also called into question existing counter-terrorism measures. Dzohkar had been a student at UMASS-Dartmouth and actually returned to campus in the aftermath of the attack.

The future may hold other terrorist attacks in store for New England, but it appears that in the aftermath of the 2013 attack, police are redoubling their efforts. Even though terrorists are capable of inflicting horrible casualties, their overall threat to America is much less than the old threat of the Soviet Union.

Instead of communist saboteurs, disgruntled workers appear to be the only real threat to many bases. Casey James Fury, a civilian worker at Portsmouth Navy Yard set a fire onboard the submarine USS Miami on May 23, 2012, so that he could get out early to spend time with his girlfriend. The intentional blaze caused $400 million in damage to the submarine.

A furious Fury was soon caught by federal investigators and sentenced to 17 years in prison, and charged with a $400 million fine. Repair work soon got underway on the Miami, but efforts were abandoned in 2013 as spiraling costs raised the price of repairs to nearly $700 million. The Miami will be decommissioned and scrapped to spare American taxpayers any further losses. The loss of Miami marks the first time that a US submarine has been destroyed by sabotage.

Today, the main US enemies abroad are small groups of self-taught terrorists and conspirators, and a few ‘rogue’ regimes like North Korea, Iran, and to a lesser extent Pakistan and Syria. But change is afoot. In light of recent growth in developing countries and the economic troubles surrounding the War on Terrorism and the 2008 financial crisis, the old Soviet Bloc is rising from the ashes of 1989.

Russia is back, with a booming economy, oil, gas, and a huge chunk of America's debt. As the Russian seizure of Crimea and continuing meddling in eastern Ukraine throughout 2014 has shown, the new, non-communist Russia is still interested in expanding its influence outside of its borders. The modern Russian Federation is a much more open and less menacing place than it once was, but it is undeniably undemocratic. Vladimir Putin's conservative United Russia party manages the country through corruption and manipulation.

The bigger long-term challenge may prove to be China. Starting in the 2010s, the Obama administration launched its 'Pivot to Asia,' repositioning US forces from the Middle East to the Pacific Ocean, to back up South Korea, Japan, and the Philippines against a Chinese military that is growing quickly. While a return to the Cold War days doesn't seem very likely, the world may be poised to become more complicated in the near future.

No matter what the circumstances are, New England is sure to be at the center of the action, supplying the brains and manpower needed to prevent—and in dire situations, fight—wars in the Middle East, Africa, Asia, or even Europe.

When and if that day comes, the uniforms may return to Pease, Quonset Point, Grenier, and Otis. Perhaps even the massive B-52s will make a comeback, touching down at Loring, placing a new generation of Aroostook County residents on the frontlines. Parking lots could fill to overflowing at Raytheon, MITRE, BAE Systems, and General Dynamics, as legions of skilled civilian workers take up the task that their parents and grandparents started in the Cold War. Outside the gates of the military bases and defense contractors, peace activists may gather once more, helping to keep militarization in check.

We must hope that the future will prove to be a sane and peaceful place, but much of the evidence seems to suggest that big challenges remain ahead. No matter the threat, New Englanders will be ready to handle it...and they will lead the charge to keep the peace.

THE END

## Index

## References


[1] "Air Force Shelves Presque Isle Base, Effective Feb. 20." *Boston Globe*, February 6, 1948.

[2] "Air Armada Set to Salute Parade in South Boston." *Boston Globe*, March 17, 1948.

[3] "Bugs, Weather Help Halt 'Invaders' at Ft. Devens." *Boston Globe*, July 17, 1949.

[4] "Grenier Field Shifted to Defensive Role." *Boston Globe*, August 24, 1949.

[5] "2 NE Air Units to be Inactivated in Cutback." *Boston Globe*, August 27, 1949.

[6] "Air Force Offers $23,000,000 Base to Presque Isle for $1." *Boston Globe*, October 4, 1949.

[7] "State Air Guard Group Trains on Mustang Fighters." *Boston Globe*, August 6, 1950.

[8] "Bostonians Training in New York to Take Over Control of Our Port." *Boston Globe*, August 13, 1950.

[9] "Old and New Seen at Exposition." *Hartford Courant*, August 15, 1950.

[10] "GI Missing in Korea Lands at Westover." *Boston Globe*, September 1, 1950.

[11] "Navy Would Quit Squantum for South Weymouth." *Boston Globe*, November 10, 1950.

[12] "100 Fighter Planes 'Attack' Facility at South Weymouth." *Boston Globe*, September 23, 1951.

[13] "Second Blast Fells Weymouth Hangar Doors." *Boston Globe*, June 7, 1951.

[14] "Air Force to Shift MATS Depot from Westover." *Boston Globe*, September 26, 1951.

[15] "NH Towns Organize Against New Air Base." *Boston Globe*, February 4, 1952.

[16] "Foreign Subs Have Been 'Sighted' Off New England Coast, Navy Says." *Boston Globe*, September 12, 1952.

[17] "250 Planes Prowl Area in 'Operation Codfish.'" *Boston Globe*, September 21, 1952.

[18] "25 Jets on Way to Patrol 'Curtain' Stop at Dow Field." *Boston Globe*, April 2, 1953.

[19] Ray, Thomas W. "A History of Texas Towers in Air Defense 1952-1964". Texas Tower Association.

[20] Rose, Jim. "Bunker N9 and Nuclear ASROC". Wompatuck News (Friends of Wompatuck State Park), 2007.

[21] "Jet Tanker Crashes; 15 Killed." *Boston Globe*, June 27, 1958.

[22] "5 Die as Jet Tanker Crashes at Maine Base." *Boston Globe*, November 25, 1958.

[23] Sanderson, Philip. "Strato-Fortress Crashes in NH Swamp; Crew Hits Silk for First Time." *Boston Globe*, August 11, 1959.

[24] "Six Sue US for Fire Loss in Jet Crash." *Boston Globe*, July 18, 1959.

[25] "Fear Explosion of Live Rockets in Crashed Jet." *Boston Globe*, October 1, 1959.

[26] "General Court Martial Denied at Ammunition Dumping Trial." *Boston Globe*, July 9, 1959.

[27] "Find Three Dud 250-Pound Bombs in Charlton Woods." *Charlton News*, February 15, 1955.

[28] "Jet Booms Startle Four Million." *Boston Globe*, September 12, 1958.

[29] "US Jet Breaks Sound Barrier and $5000 Worth of Windows." *Boston Globe*, August 7, 1959.

[30] Kline, Nat. "Supersonic Air Team at Otis AFB Thursday." *Boston Globe*, August 9, 1959.

[31] "Shah of Iran Flies to Hub Before Trip to Visit Ike." *Boston Globe*, June 30, 1958.

[32] Forbes, Donald H. *Two Communist Brothers from Washington, New Hampshire and their Fight Against Fascism.* 2013. Print.

[33] "Graduate of School at Sub Base Made Record Dive in Bathyscaph." *Hartford Courant*, January 26, 1960.

[34] "So. Weymouth Pilots in Big ASW Exercise." *Boston Globe*, April 17, 1960.

[35] "Air Ground Pickup Line Tested for Arctic Rescue." *Boston Globe,* August 10, 1960.

[36] 'Operation Orphans' Airplane Full for Flight to Korea." *Boston Globe*, December 4, 1960.

[37] "250 Reservists Called to Ferry C-119s from France." *Boston Globe*, January 31, 1961.

[38] "Buildings Shook, Babies Wailed." *Boston Globe*, February 20, 1961.

[39] McGrath, Edward. "NE Reservists see Navy Help Cubans Fight Polio." *Boston Globe*, March 8, 1961.

[40] Kline, Nat L. "First Nuclear Frigate to be Launched Here." *Boston Globe*, April 9, 1961.

[41] "B-52 Refuels Sky Tanker in Air to Avert Crash." *Boston Globe*, August 12, 1961.

42 "NE Units in Paradrop." *Boston Globe*, August 12, 1961.
43 "Thresher Returns Home After Shakedown Cruise." *Hartford Courant*, September 9, 1961.
44 "Largest Radome in World Built at Tyngsboro." *Boston Globe*, September 17, 1961.
45 "Air War Rages." *Boston Globe*, October 16, 1961.
46 Sumner, Barton. "Automatic Weather Recording for Jets, Missiles." *Boston Globe*, December 17, 1961.
47 "Roscoe Ammon, Manchester N.H., Industrialist." *Boston Globe*, December 17, 1961.
48 "80 Pct. of SAC Alert on Holiday." *Boston Globe*, December 26, 1961.
49 "Near Miss Mars Armed Forces Day: Planes Escape Hanscom Crash." *Boston Globe*, May 20, 1962.
50 Menzies, Ian. "Little Hope for Living After Nuclear Attack." *Boston Globe*, May 31, 1962.
51 "7 Southern Riders in NH Land Jobs." *Boston Globe,* July 12, 1962.
52 "Three Crewmen Die in Tanker Crash." *Hartford Courant*, August 9, 1962.
53 "Fires of Missiles Worry Fire Chiefs." *Boston Globe*, June 5, 1962.
54 "US Files Suit Against 3 Milk Cos." *Hartford Courant*, August 11, 1962.
55 "Bombers Take Over." *Boston Globe*, October 24, 1962.
56 "Dow AF Base Wins War With Birds, Insects." *Boston Globe*, October 6, 1962.
57 "Picketers Plan for Launching." *Hartford Courant,* January 11, 1963.
58 "B-52 Down in Maine, 9 on Board." *Boston Globe*, January 25, 1963.
59 "Halifax Home for Red Sailors." Boston Globe, Aug. 4, 1963.
60 "Halifax Home for Red Sailors." Boston Globe, Aug. 4, 1963.
61 "Hearing Set in DC on Red Fishermen." Boston Globe, Sept. 1, 1963.
62 "Pentagon Calm at Red Ships." Boston Globe, Sept. 1 1963.
63 Banner, Earl. "N.E. Fisheries Ruin Charged as Red Aim." Boston Globe, Sept. 6, 1963.
64 "B-52 Carrying A-Bombs Crashes, No Peril Seen." *Boston Globe*, January 14, 1964.
65 Rodgers, Wilfrid. "NASA Picks Boston for Space Center." *Boston Globe*, February 1, 1964.
66 "Giant Electronic Eye is Nearing Completion." *Hartford Courant*, March 30, 1964.
67 "US Fights for Boat 'Battle of Groton' Prize." *Hartford Courant*, May 22, 1964.
68 "Pilot Dies Trying to Save 150." *Boston Globe*, July 19, 1964.
69 McGrath, Edward. "Red Trawlers Off Cape Called Part of Russian Spy Fleet." Boston Globe, Sept. 30, 1964.
70 Wise, David. "McNamara Dooms 95 Bases." *Boston Globe*, November 20, 1964.
71 "Shipyard in NH Told No Cuts for 3-4 Years." *Boston Globe*, November 27, 1964.
72 "Dow Air Base Closing Brings Bangor Protest." *Boston Globe*, December 2, 1964.
73 Rollins, Bryant. "Let's Have a Little Talk Jan. 25." *Boston Globe*, December 10, 1964.
74 "Perini Wins Contract for AF Missile Bases." *Boston Globe*, February 26, 1965.
75 Kline, Nat. "1-Man Moon Staff Grows at Hanscom." *Boston Globe*, March 28, 1965.
76 "Boston Army Base Doomed by 1970." *Boston Globe*, December 9, 1965.
77 Banner, Earl. "Fishermen Want to Know What Bureau is Doing." Boston Globe, March 8, 1965.
78 Beatrice, Mike. "U.S. to Check Soviet Fishing Vessels." Boston Globe, April 26, 1965.
79 Banner, Earl. "Red Fleet Offers Aid to US Fishermen." Boston Globe, May 24, 1965.
80 Banner, Earl. "USSR Plans Sea Lab Off N.E. Coast." Boston Globe, November 16, 1966.
81 Banner, Earl. "Russian Ship is Coming Sunday." Boston Globe, August 17, 1967.
82 Godsoe, William. "Arms Day Programs Cut by War." *Boston Globe*, May 19, 1966.
83 "UFO Over Boston Area." *Boston Globe*, May 22, 1966.
84 "Spacecraft in Hub Today." *Boston Globe*, June 8, 1966.
85 Godsoe, William. "New Radio Unit Guides F-102." *Boston Globe*, August 20, 1966.
86 "Hangar Coming Down." *Boston Globe*, September 3, 1966.

[87] "Anti-Sub Warfare Exercises Set for Narragansett Bay." *Boston Globe*, September 18, 1966.
[88] "Brazil Students to Visit Base." *Hartford Courant*, June 15, 1967.
[89] "Navy Probes Sloop, Sub Crash." *Hartford Courant*, September 6, 1967.
[90] "Navy Pollutes River Despite State Pleas." *Hartford Courant*, November 28, 1967.
[91] Collins, James. "Boy Flees Operating Table—and Keeps Tonsils." *Boston Globe*, November 7, 1967.
[92] "Groton Radiation Linked to N-Subs." *Hartford Courant*, March 20, 1968.
[93] "Navy Retention Problems Blamed on Poor Housing." *Hartford Courant*, June 13, 1968.
[94] Riley, Arthur. "Science in Industry: MITRE: it helps guard free world." *Boston Globe*, October 13, 1968.
[95] "Apollo Crew's Hot Meals Made in Natick." *Boston Globe*, October 13, 1968.
[96] "Natick's Tunnel Explorer." *Boston Globe*, September 22, 1968.
[97] "Maine Jet Tanker Crashes in Pacific, 11 Die." *Boston Globe*, September 25, 1968.
[98] "2 From Mass. in SAC Crash Over Taiwan." *Boston Globe*, October 24, 1968.
[99] McClintock, Ralph. "My Pueblo Nightmare." *Boston Globe*, May 16, 1969.
[100] "Boston's Fishing Industry Perks Up." *Boston Globe*, March 31, 1968.
[101] Caldwell, Jean. "Westover AFB to Lose SAC HQ." *Boston Globe*, August 26, 1969.
[102] "Naval Air Station in So. Weymouth May Move to Quonset Next Year." *Boston Globe*, December 12, 1969.
[103] Healy, Robert. "Was it wrong to lower the flag?" *Boston Globe*, May 8, 1970.
[104] "Fires, Violence Flare for 2d Day in New Bedford's Black Section." *Boston Globe*, July 10, 1970.
[105] "Street riots can be ended." *Boston Globe*, July 11, 1970.
[106] Sheehan, Alan. "Mayor Clamps Curfew to Calm New Bedford." *Boston Globe*, July 13, 1970.
[107] Wood, John B. "1500 Antiwar protestors in Boston march to the Navy Yard." *Boston Globe*, May 14, 1972.
[108] "At Smith's commencement...flowers, protests, tears." *Boston Globe*, May 29, 1972.
[109] Nugent, Rory. *Down at the Docks*. New York: Pantheon Books, 2009. Print.
[110] Dunlop, Tom. "The Defection of Simas Kudirka." *Martha's Vineyard Magazine*, 2005.
[111] Woodard, Colin. *The Lobster Coast: Rebels, Rusticators and the Struggle for a Forgotten Frontier.* New York: Penguin Books, 2005.
[112] Unger, Sanford J. "Pilot refuses raid on Cambodia." *Boston Globe*, July 5, 1973.
[113] "Federal judge rules US must halt bombing." *Boston Globe*, July 26, 1973.
[114] Anderson, Peter. "Fresh troops to bring peace to Telegraph Hill." *Boston Globe*, October 11, 1974.
[115] Nyhan, David. "Ford declines to send US troops to Boston." *Boston Globe*, October 16, 1974.
[116] "White supremacy party commander arrested during Boston busing rally." *Boston Globe*, October 27, 1974.
[117] Wilkie, Curtis. "Hundreds of police quell disturbances at Carson Beach." *Boston Globe*, August 11, 1975.
[118] Kenney, Michael. "600 Guardsmen Moved to City; protests follow." *Boston Globe*, September 8, 1975.
[119] Rogers, David. "2 injured as demonstrators confront guardsmen in S. Boston." *Boston Globe*, September 8, 1975.
[120] Anglin, Robert. "He did his job and was blinded." *Boston Globe*, February 18, 1976.
[121] "Bombs found at schools." *Boston Globe*, October 26, 1976.
[122] "Soviet TV on Boston." *Boston Globe*, April 19, 1976.
[123] Satchell, Mitchell. "UFO's vs. USAF: Amazing (but true) Encounters." *Boston Globe*, December 10, 1978.
[124] Patterson, Rachelle. "Bases in New England may get a shot in the arm." *Boston Globe*, November 3, 1977.
[125] Stewart, Richard. "Nine hours in a B-52: Life aboard a Maine-based strategic bomber." *Boston Globe*, June 12, 1977.

[126] "Iranian Sailors Briefed; Base Tightens Security." *Hartford Courant*, February 18, 1979.
[127] Vaughan, Kristi. "Atom Personnel Leaving the Navy." *Hartford Courant*, December 7, 1976.
[128] Whalen, Michael J. "Navy Offers Drug Therapy to N-Submarine Crews." *Hartford Courant*, March 10, 1978.
[129] "Grounding of Sub Reported." *Hartford Courant,* June 21, 1979.
[130] "Bomb Threat Proves Hoax." *Hartford Courant*, April 26, 1975.
[131] Roy, Suzanne. "400 Naval Personnel Evacuated During Fire at Submarine Base." *Hartford Courant*, July 14, 1980.
[132] "FB111 Explodes in Maine Woods." *Boston Globe*, December 24, 1975.
[133] Roberts, Fletcher. "Jet Crash Off of Vineyard Kills Pilot." *Boston Globe*, April 6, 1978.
[134] Estes, Andrea. "Soldier Convicted in Slaying." *Boston Globe*, February 8, 1980.
[135] "Ex-Soldier is Free in 1977 Rape Case." *Boston Globe*, December 29, 1982.
[136] "A Green Beret accused of threatening to kill." *Boston Globe*, August 31, 1983.
[137] Zeitlin, Arnold. "At Pease AFB, Nuclear Bombs and Guards on Drugs." *Boston Globe*, August 6, 1980.
[138] Schneider, Andrew. "Civil Defense in a Shambles." *Boston Globe*, April 20, 1980.
[139] Ellement, John. "Parachute Fails to Open; Woman Falls to Death." *Boston Globe*, October 20, 1980.
[140] Murphy, Jeremiah. "Freedom of Speech Works Two Ways, General." *Boston Globe*, November 21, 1980.
[141] Stewart, Richard H. "No Water, No Students at UMASS." *Boston Globe*, September 5, 1980.
[142] "Soviet Planes Tracked 200 Miles Off the Cape." *Boston Globe*, November 14, 1980.
[143] Mahoney, Frank. "More is Going on at Camp Edwards Than Meets the Eye." *Boston Globe*, November 27, 1980.
[144] "Voluntary Ammo Search Set..." *Boston Globe*, January 22, 1981.
[145] Foreman, Judy. "4-Day Protest Planned Against the Euromissiles." *Boston Globe,* October 20, 1983.
[146] Higgins, Richard S. "Portsmouth Home of F-16s for Now; Residents Near Base Debate Reagan Action," *Boston Globe,* July 28, 1981.
[147] "New England Briefs; Quakers vs. Jets." *Boston Globe*, August 31, 1981.
[148] Stanton, Mike. "Westover in Thick of Air War Game." *Boston Globe*, August 23, 1981.
[149] Hebert, Josef. "Soviet Airline Curbed After N.E. Overflights." *Boston Globe,* November 17, 1981.
[150] Quinlan, Joe. "Chicopee Marine is Buried." *Boston Globe*, November 13, 1983.
[151] Kaufman, Jonathon. "Derailment Fire in Greenfield." *Boston Globe,* March 30, 1985.
[152] "51 Protestors Arrested at Cape Cod Base." *Boston Globe*, June 17, 1985.
[153] "US Protests Mark Anniversary of Nagasaki Bomb." *Boston Globe*, August 10, 1985.
[154] Hanafin, Teresa M. "Falmouth Wells May Not be Safe for a Year." *Boston Globe*, December 1, 1985.
[155] "Devens Drug Probe Results in 5 Arrests." *Boston Globe*, July 24, 1986.
[156] Hanafin, Teresa M. "State to Alter Scope of Cape Cancer Study." *Boston Globe*, August 10, 1986.
[157] "Air Force Measures Microwaves on Cape." *Boston Globe*, September 24, 1986.
[158] Ribadeneiro, Diego. "Otis to Get Fleet of F-15s to Replace Aging F-106s." *Boston Globe*, January 6, 1987.
[159] Tye, Larry. "Seabrook Eyes Fort Devens for Accident Evacuation." *Boston Globe*, September 19, 1987.
[160] "Cohen Says Brunswick, Loring Safe." *Boston Globe*, December 25, 1988.
[161] Bickelhaupt, Susan. "Ft. Devens Could be Bigger Than Ever." *Boston Globe*, December 30, 1988.
[162] "Fourth Body Found in Canada Plane Crash." *Boston Globe*, October 6, 1989.

Made in the USA
Monee, IL
08 November 2026

67969858R00118